Praise for
Chaser

"There seems to be no end to Chaser's abilities . . . [Pilley's] approach has led to a new understanding of canine intelligence, one that makes us wonder just how sophisticated a dog's mind can be." — *USA Today*

"If a truly great book leaves one better for having read it, then *Chaser* is quite simply a masterpiece. Dogs and those of us who love them owe a debt of gratitude to the brilliant, courageous author and his equally heroic subject." — Jennifer Arnold, author of *Through a Dog's Eyes*

"Chaser is the most scientifically important dog in over a century. Her fascinating story reveals just how sophisticated a dog's mind can be." — Brian Hare, coauthor of *The Genius of Dogs*

"This is an extraordinary book, full of warmth and wisdom that has the potential to forever change the way we look at dogs." — Jeffrey Moussaieff Masson, author of *When Elephants Weep* and *Dogs Never Lie About Love*

"After you read *Chaser,* you will realize that you may have underestimated the intelligence of your dog. Marvelous insights into a dog's mind." — Temple Grandin, author of *Animals in Translation* and *Animals Make Us Human*

"A delightful memoir that offers a challenge to behavioral psychologists and inspiration for pet lovers." — *Kirkus Reviews*

"This marvelous blend of good science and heartwarming dog story will inspire all of us to reexamine our canine friends." — *Booklist*, starred review

Chaser

Unlocking the Genius
of the Dog Who Knows
a Thousand Words

JOHN W. PILLEY
WITH Hilary Hinzmann

Mariner Books
Houghton Mifflin Harcourt
Boston • New York

First Mariner Books edition 2014

Copyright © 2013 by John W. Pilley, Jr.

For information about permission to reproduce selections from this book, write to Permissions, Houghton Mifflin Harcourt Publishing Company, 215 Park Avenue South, New York, New York 10003.

www.hmhco.com

Library of Congress Cataloging-in-Publication Data
Pilley, John W.
Chaser : unlocking the genius of the dog who knows a thousand words / John W. Pilley with Hilary Hinzmann.
pages cm
ISBN 978-0-544-10257-6 (hardback)   ISBN 978-0-544-33459-5 (pbk.)
1. Chaser (Border collie) 2. Dogs — Psychology. 3. Dogs — Training.
4. Human-animal communication. I. Title.
QL795.D6P55 2013
636.737'4 — dc23
2013026340

Printed in the United States of America
DOC 10 9 8 7 6 5

To Sally

Contents

There are some simple truths . . . and the dogs know what they are.

— Joseph Duemer, *A Dog's Book of Truths*

Chaser

1

The Dog Who Knows
a Thousand Words

THE TWENTY OR so nine-year-old boys and girls squirmed and murmured in their seats as Chaser walked into their fourth grade classroom at Public School 31 in Brooklyn.

"She's here!"

"She looks bigger on TV."

"She's so cute!"

The children's desks were pushed to the back and sides of the room, positioned so they all had a good view of the space that had been cleared at the front of the classroom.

I let Chaser off the leash and she went straight to my grandson, Aidan, who was sitting at one end of the first row with a giant smile plastered on his face.

Aidan had been begging us to bring Chaser to visit his class. He had told his teacher, Mrs. Tapper, about our visit to New York for several national television appearances, and most of the class had watched Chaser demonstrate her unprecedented language skills on *Nova scienceNow* a few days before. Now here we were, about to present those skills in person.

After greeting Aidan, Chaser moved on to the other children, looking up at each one expectantly and wagging her tail. As Aidan's classmates leaned down to pet Chaser, her tail wagged harder and her tongue lolled out of her mouth in an ecstatic canine grin.

Chaser is not an imposing dog physically. She's normal Border collie size, about twenty inches tall at the shoulder and about forty pounds. Her thick, fluffy coat is mostly white with splotches of gray and flecks of black, except for large patches of black on the left side of her head, both sides of her torso, and her hind legs. Although she has plenty of spirit, overall she has a soft temperament. I often tell people, "She's a lover, not a fighter."

All the same, a few children were a little shy about touching Chaser. Any dog can be intimidating for people who aren't used to them. Chaser has lots of experience with children, however. She knows that they're the best possible candidates for her favorite activity, play. She wiggled and squirmed and wagged her tail appealingly in front of the shier kids, and soon even the most reluctant children were grinning right back at her.

It was time to begin. We only had an hour and there was a lot to show. The principal, Mrs. Scarlato, went and stood in the back of the classroom with her camera, and Mrs. Tapper introduced my wife, Sally, our daughter (and Aidan's mom) Debbie, and me to the class. Then I called Chaser to me and she lay down at my feet.

"I'm Dr. Pilley," I told the children, "and I'm a scientist. Can anyone tell me what scientists do?"

One boy's hand shot up, and I pointed to him. "Scientists invent things," he said. Another offered, "They look at stars and rocks and stuff like that." And a girl said, "They study plants and animals, too."

I asked, "Why do scientists study things like rocks and stars and oceans and plants and animals? What are they looking for?"

Another girl said, "They want to discover things."

"Aha, that's it! That's what we scientists are after, discovery! Chaser and I have been trying to discover how much human language she can learn." I see Chaser as a co-investigator and research assistant rather than as an experimental subject. Just as she's a part of our family, she's also the other half of my research team. "I understand you've heard that Chaser knows more than a thousand words."

Over the previous few weeks there had been headlines around the world about Chaser. It all began just before Christmas, with the e-publication of a peer-reviewed scientific article on my research with Chaser. I'd written it with the help of Alliston Reid, a former student of mine at Wofford College in Spartanburg, South Carolina, where he had succeeded me as a professor of psychology.

The article, published first online and later in print by the British journal *Behavioural Processes,* reported that Chaser had learned and retained the proper noun names of 1,022 objects over a period of three years. She had also learned to understand and distinguish the separate meanings of proper noun names and commands, and had learned at least three common nouns, or words that represent whole categories of things. Furthermore, she could learn a new word by exclusion, meaning that she could infer the relationship between a name she had never heard before and an object she had never seen before, by picking the new object out of a group of objects whose names she already knew. These abilities are normally seen in children when they are acquiring language as toddlers, and there is fierce debate among scientists as to whether animals can really do the same things.

On December 8, 2010, the day the article was published online, I received a phone call from a journalist for the British magazine the *New Scientist.* Two weeks later, the online edition of the *New Scientist* ran a story with the headline "Border Collie Takes Record for Biggest Vocabulary." Our phone started ringing off the hook and my e-mail inbox began filling up with messages from print, online, and broadcast media. Over the week from Christmas to New Year's the Chaser story "went viral," as Debbie's husband, Jay, explained it to me.

In more than forty-six languages, the headlines ran the gamut from silly to serious. The *National Examiner* put Chaser on the cover next to Brad Pitt and Charlie Sheen. Most of the media attention centered on Chaser's knowing more than a thousand words, which MSNBC noted is more than most toddlers know. There was also lots of speculation about whether Chaser was "the world's smartest dog."

"So far as it's been documented," I told the class, "Chaser knows more words than any other dog. In fact, she knows more words than any other animal except humans. We're going to demonstrate Chaser's learning of words for you in a minute. But first I want to show you some learning that she shares with a lot of Border collies."

I opened the plastic bin we'd brought with us and showed the children what was inside: twenty of Chaser's toys, mostly stuffed animals and dolls, but also balls and Frisbees (our word for any spinning disk toy). Each one had a name written on it in permanent ink, like "Bamboozel," "Frat Rat," "Bear," "Crock," "SpongeBob," "ABC," "Prancer," and "Santa Claus." Among the rest of her toys at home there was more than one teddy bear, but the other bears had unique names with no repeats. That made it possible for Chaser to keep them straight in her head.

I put one of the toys, a foam Frisbee with "Flipflopper" written on it, down on the floor about eight feet away from Chaser. She had her head down as if she had no interest in what I was doing. Only experience told me that she was "giving an eye to the sheep and an ear to the farmer," as Border collie trainers say. Flipflopper was one of Chaser's flock of named toys, one of the 1,022 surrogate sheep she has acquired in her language training. Any activity to do with Flipflopper was important to her.

With her ears down and her eyes half closed, Chaser looked like she was in a trance state, as more than one journalist has described her in these moments. But she had her legs and feet drawn under her evenly on both sides. That was the one telltale sign that she was ready to spring into action as soon as I spoke the words that told her what I wanted her to do.

"Chaser, go out," I said, using the words that sheep farmers have traditionally used to start their dogs toward some sheep.

Chaser instantly got to her feet and walked toward the foam Frisbee.

"There," I said, and Chaser stopped still.

"Chaser, come by," I said, and she walked around the Frisbee clockwise.

"There," I said, and she stopped again.

"Chaser, way to me," I said, and she walked around the Frisbee counterclockwise.

The children were murmuring with interest and inching forward in their chairs.

"There," I said, and Chaser stopped still again.

"Chaser, drop." She dropped flat on her belly. Some of the kids started giggling.

"Chaser, crawl," I said, and she began to crawl toward the Frisbee. More giggling from the kids.

"Chaser, one, two, three, take!" I said, my special signal to release her to rush to the toy and take it in her mouth. She began shaking the Frisbee with delight, triggering applause from the kids, and I encouraged her in that for ten or fifteen seconds. Then I said, "Chaser, bring Flipflopper. Drop Flipflopper in the bin."

When she'd dropped Flipflopper in the bin I told the class, "That's Chaser's reward. She gets to play with the toy. On a sheep farm, a Border collie's reward would be moving the sheep around. Border collies have a strong herding instinct, and that's what they like to do best. Now, for Chaser, each of these toys represents a sheep that she gets to herd in different ways. Like I said, she knows more than a thousand objects by name. The next closest dog, a Border collie in Germany named Rico, knew a couple of hundred words, and Rico was part of my inspiration for the scientific study Chaser and I have done. Now we'll demonstrate Chaser's word learning for you."

I had written the names of four of the toys in the bin on note cards, and I gave the cards to two of the boys and two of the girls.

"Don't call out what's on your card yet," I told them. "First I'll ask Chaser to find some toys by name. And then I'll ask the four of you to ask Chaser to find the toys whose names are written on the cards."

I dumped all the toys out of the bin in a jumble on the floor several feet away from Chaser. She again had her head down as if she had no interest in what I was doing. I came back over to her and said, "Chaser, find SpongeBob."

Chaser sprang to her feet and trotted over to the toys. She looked over the toys for a moment, and then spotted SpongeBob at the bottom of the pile and pulled him out. That really excited the kids.

"Shake SpongeBob, Chaser," I said, to give her a moment's play as reward, and then had her bring SpongeBob over and drop him in the bin.

I asked Chaser to find and bring over nine more toys, calling for each by its unique name. With ten toys still left on the floor, but spread out a bit now by Chaser's nosing and pawing among them, I asked each of the four students in turn to ask Chaser to find the toys written on the note cards.

"Chaser, find Santa Claus," the first child said.

"Chaser, find Bamboozel," the second child said.

"Chaser, find Mickey Mouse," the third child said.

"Chaser, find Croc," the fourth child said.

One by one Chaser found the toys. After she'd had a chance to play with a toy for a few moments, which always delighted her, I asked her to deposit it in the bin.

Each time Chaser did something one of their classmates asked for, the class got more and more excited. It was a shame we couldn't give them all a chance to work with her. But in the time we had left, I wanted to show the kids two more aspects of Chaser's language learning.

Earlier I had asked the kids how many of them had a dog at home, and about half did. When I do classroom demonstrations in places where most people live in houses with backyards, practically every kid will raise their hand.

Now I said, "Those of you who have dogs, when you tell your dog, 'It's time to go for a walk,' does your dog understand that each of those words has a separate meaning?"

They all said, "No."

"What word do you think your dog does know?"

The kids gave various answers, but the consensus was that their dogs all knew the word *walk*.

"Well, 'It's time to go for a walk' is a pretty long sentence. How about something simpler? If you throw a ball and say, 'Fetch ball,' does your dog understand that those two words have separate meanings?"

This was a tough question, but the kids finally decided that their dogs really only understood the "fetch" part.

"Many scientists think that's all a dog can ever learn. They say that dogs don't really understood that different words have separate meanings; they just learn to associate your words and gestures with certain actions you want them to make, like 'sit,' 'stay,' or 'fetch.' Chaser and I did another experiment to test exactly that."

I took three of Chaser's toys, Bear, Croc, and Prancer, and put them on the floor.

"Okay, class, we have three toys, and we can use three different commands. We can ask Chaser to nose a toy, paw it, or take it, which means take it in her mouth. And we can mix the three commands up with the three objects any way we like. Let me show you."

In succession, I asked Chaser to nose, paw, and take different toys, and then I picked a couple of students to do the same.

"What do you think, class?" I asked. "If we can use different commands with different toys, does this show that Chaser understands that a word for an action means one thing and a word for an object means another thing?"

"Yes!"

"I agree with you," I said. I didn't add that I was waiting to see if any scientists in the fields of animal learning or childhood language learning would find fault with my experiment and cast doubt on its results. As it happened, no one did. In fact, in the three years since the *Behavioural Processes* paper was published, no one has rebutted any of its specific findings, although the controversy about animal learning continues unabated. Meanwhile, Chaser's learning has also continued, reaching remarkable new heights that include adding many common nouns to her vocabulary, understanding sentences with three elements of grammar, and progress in learning by direct imitation of my actions, as I'll explain later.

"I wish Chaser and I could stay longer with you," I told the class, "but we're running out of time, and there's one more thing we really want to show you. Chaser can learn the name of a new object even though she's never seen the object or heard its name before, and she can do it in one try. All the toys I brought with me are things Chaser knows the name of, so we need a new object. But we don't want Chaser to see it yet. Aidan, why don't you take Chaser behind Mrs. Tapper's desk? That way we can be sure she can't see anything."

Aidan grinned at his classmates, looked at Mrs. Tapper to make sure she had no objection, and took Chaser behind the desk and crouched down with her. For good measure he put his hands over Chaser's eyes, and I thanked him for doing that.

Then I asked the class, "Does anyone have an object we can use?"

Lots of kids raised their hands. But a girl in the second row caught my eye by waving a bright red plastic change purse.

"That looks good," I said, taking the purse from her outstretched hand. "What are we going to call it?"

"Purse!" several kids said.

"We could do that, but Chaser's heard the word *purse* many times, and she probably knows that Aidan's mother and grandmother and other people have purses. So she might already have an association for the word *purse* in her mind. We want a name that will be completely new to Chaser. It could be anything, it could be just a silly sound, like 'Woosh.' Should we call this 'Woosh'?"

This sparked smiles and laughs. "Yeah, let's call it 'Woosh,'" the kids agreed.

I said, "We'll put down some objects that Chaser already knows the names of, and then we'll add this new one. She doesn't know what it looks like, and until just now she's never heard the special name we're giving it.

"Before we see if Chaser can find the new object when we ask for it by its special name, let's ask Chaser to find some familiar objects. She might come over here and take the new object just because it is new and she's curious about it. We all like to get new things to play with, don't we?"

I didn't tell the children that a well-known childhood language learning researcher had criticized a study of word learning by the Border collie Rico precisely because it had not established whether the dog had "a baseline preference for novelty" when it came to learning the name of a new object. This was one of a number of criticisms by prominent researchers of the Rico study, which was published around the same time Sally and I got Chaser as a little puppy. I had carefully designed my own experiments, such as asking Chaser to find some familiar objects first, in the hope of avoiding such criticism.

At this point, I asked Aidan to go back to his seat and I called Chaser over to me. Then I said, "Chaser, find Mongrel." Chaser found Mongrel among the toys. And then I asked her to find other familiar toys by the names she'd learned for them. She found them all without a glitch, ignoring the unfamiliar object and demonstrating that she had no baseline preference for novelty. After each trial with a familiar toy, we put it back in the pile and jumbled all the toys up again.

Finally I said, "Chaser, find Woosh. Find Woosh."

Chaser went over to the toys and looked them over carefully. She pawed them a bit, and then bent down and picked up the red plastic coin purse.

"Yes! Good girl, Chaser! Good girl! Chaser, bring Woosh. Bring Woosh."

She brought the red purse over to me and dropped it in the bin.

The kids loved that, and they gave Chaser their biggest cheers. She responded with body wiggles and tail wags, ears up, eyes wide open, tongue lolling out of her mouth — all signs of how pleased she was to win the kids' attention and affection. Then I said it was unfortunately time for us to leave and suggested that everyone come to the front of the room and have their picture taken with Chaser. Chaos erupted, and the children crowded so tightly around Chaser that she couldn't be seen. She didn't mind. She loved being petted and stroked by the children. But the chaos had to end, and Mrs. Tapper and Mrs. Scarlato good-humoredly restored order and arranged the children around Chaser for a photograph.

My last words to the class were to remind them that play was Chaser's reward. I told them that they should always reward their dogs for good behavior by playing with them and petting and praising them. We all learn better and faster when learning is fun.

A couple of days later, before Sally and I drove home to South Carolina with Chaser, Aidan came home from school with giant thank-you cards that he and his classmates had made, each with a drawing of Chaser and a signature. One little boy put himself in the picture with Chaser, giving himself a big red heart for a body and big stars for eyes. It was such an eloquent way of saying that he loved Chaser and she was a star in his eyes. A little girl drew a smiling Chaser with her toys, giving her a gold "Chaser" nametag on a pink collar that matched the pink of the insides of her ears. It was really touching to receive all of the children's drawings.

Throughout the classroom demonstration, Chaser had been in her element. Finding objects, herding them, learning new objects and names, interacting with the children — in all these things Chaser was expressing her intensely social nature, a characteristic that all domestic dogs share in different ways, as well as her strong Border collie instincts and drives. Her unprecedented language learning rests on these two factors and on the relationship that Sally and I have built with her around them.

Together the three of us have gone on a journey of discovery we could never have anticipated. I had been retired for eight years when Chaser came into our family and reawakened my passion for discovery — really, reawakened me as a person as well as a scientist. But Chaser was not my first Border collie, and not the first dog to be my co-investigator and research assistant. Twenty years before Chaser, there was Yasha.

2

Goodbye

I CANNOT TELL CHASER'S story without first telling Yasha's story. Yasha was as pivotal as Chaser in my efforts to understand canine intelligence. A brilliant and adventurous Border collie–German shepherd mix, Yasha was — outside of Sally, Debbie, and our other daughter, Robin — my best friend for sixteen years. Where I went, Yasha went. A faithful companion, he taught me infinitely more than I taught him.

Yasha joined the Pilley family late in the spring of 1978, taking up residence in our two-story, three-bedroom home in Spartanburg, South Carolina, just a few miles from Wofford College, where I'd been a psychology professor since 1969. Bimbo, our big, floppy-eared red and brown German shepherd mix, had recently passed away from old age and cancer. Rough-and-tumble if need be, but a sweetheart in the family, Bimbo had ridden shotgun on Robin and Deb's horseback riding adventures growing up, and we all missed him. The girls were settling into summer jobs after their latest college semesters, and they complained that things weren't the same without a dog to liven things up. We'd always had a dog in the family, and both girls were begging us to get a new puppy.

Sally knew I'd become more and more intrigued by Border collies since I'd met the local breeder and trainer Wayne West a few years earlier. Their problem-solving ability and receptiveness to complex sequences of verbal commands fascinated me.

When she was at the office of Rice McFee, our longtime vet, writing out a check to pay off the bill for Bimbo's care, she half scolded, half teased him: "You haven't forgotten about us, have you? Remember, you have to tell us if you know of any good puppies being available."

Rice McFee's face went blank for a second, then lit up with a huge smile. "As a matter of fact, Doug Chappell's Border collie just had a litter of puppies. They are a mixed breed, so he's not sure who the daddy is, but I understand a neighbor's German shepherd is the likeliest candidate, or culprit," he said with another smile. "In any case, they are ready to be weaned, and I bet you can go over right now."

When I came home a couple of hours later, I found Robin and Deb playing with a little brown and white ball of fur with legs. My entrance immediately attracted the puppy's attention. Tail wagging, he trotted over to me to say hello, and I knelt down to pet him. I lifted him up and held him close to my face, responding to his face licks with kisses and soft coos.

Robin said, "His name should be Jascha, Dad."

Debbie chimed in, "We're already calling him that, so you will just confuse him if you change it."

Jascha, the girls explained, was for Jascha Heifetz, whose recording of Mendelssohn's violin concerto was a family favorite. But I have a tendency to misspell names, especially when it comes to our family pets. "He looks like a Yasha," I said, and the name stuck.

Sally sidled up to give me a hug. With a grin she asked, "What do you think?"

"I think he's wonderful. You chose the pick of the litter, babe," I said, giving her a squeeze.

Although we all wanted the puppy, I was the one with the most time to devote to him over the next few weeks. Sally had to be at the hospital early every morning for her head nurse shift in the critical care department, and in addition to starting their summer jobs on the right foot, the girls were busy reconnecting with old friends. Rather than leave the puppy alone for long stretches, I took him with me to my lab at Wofford.

Eight-week-old Yasha was the brightest puppy I had ever encountered. He picked up the basic obedience commands as if he'd always known them. He was also eager to engage with people and quickly made himself at ease with strangers. The dean of Wofford at that time had an old hound dog who went anywhere he pleased, and he became Yasha's canine guide to the campus.

A few weeks after Yasha joined the family, Sally and I were due to leave on a three-week trip to Eastern Europe. The timing was unfortunate. Neither of us wanted to tear ourselves away from our new puppy. As our departure date approached, I started prepping the girls with work that I wanted them to do with Yasha while Sally and I were gone. Robin and Debbie were sure to take good care of him, but I didn't want his training to lapse. I gave the girls strict instructions to continue his early obedience work and teach him something new every day.

The trip behind what was still the Iron Curtain was incredible. But it was great to get home to Spartanburg and reunite with Robin, Debbie, and Yasha. I didn't waste any time in asking the girls, "Did you teach Yasha something new every day?"

Brunette, hazel-eyed Robin and blond, blue-eyed Deb shared a smirk, displaying the high cheekbones they inherited from Sally and the dimples they got from my mother, and replied, "We taught him a lot!"

The girls gave us a performance of Yasha's new repertoire. Most of their teaching ended up being what David Letterman refers to as "stupid pet tricks." Yasha was so eager and fast to learn that the girls quickly started teaching undignified tasks such as "crawl," "grovel," "get that flea," "cry," "get your tail," and "sneeze."

"Very impressive," I said dryly. Sally chuckled and I had to repress my own laugh. In hindsight, I should have been more specific about the obedience training I wanted Robin and Debbie to give Yasha. In elementary school the girls' usual afterschool routine meant coming to my lab, where they did their homework and played with the rats and pigeons until it was time to go home for supper. They'd both done animal training experiments for science projects. On top of that, Robin had become a psychology major, and she was well versed in operant conditioning, which essentially means finding a behavior the animal already does, naming it, and then reinforcing it, thus bringing it under "stimulus control." So Robin and Deb had caught Yasha in the middle of sneezing or scratching and then named and reinforced the behavior so that he now did it on command.

Robin knelt down in front of Yasha. Leaning forward slightly as she spoke, she said, "Yasha, what's two plus two?"

Yasha quickly gave four sharp barks. I said, "Nice — so he knows how to bark to four."

Robin ignored me and spoke to him again, very slowly and deliberately. Yasha sat in front of her, his ears at attention, as she said, "Yasha, what's five plus two?" This time, seven sharp barks answered. Robin ran through several other little sums, and Yasha gave the correct number of barks every time.

I smiled widely and said, "Impressive. How are you doing it?" I knew this was an example of a Clever Hans effect, an issue in all animal learning experiments and something I'll explore later. Robin teased me for a bit, declaring that she had successfully taught Yasha the rules of addition and was going to teach him subtraction next. Finally she fessed up that she leaned forward a little to cue Yasha that the command to speak — "Yasha, what's . . ." — was coming. When she wanted him to stop barking, she leaned back and said, "Good dog!" Yasha didn't really know how to add. He simply read Robin's body language and barked until she cued him to stop. His skill at arithmetic always entertained the girls' friends, but it wasn't on my shortlist to teach my new puppy.

I grumbled a bit about Robin and Debbie's not having taught Yasha useful things such as the American Kennel Club (AKC) obedience exercises. I'd started him on the novice exercises before going to Eastern Europe. But the girls' mischievousness showed that Yasha had a flair for learning new behaviors. That was intriguing.

Genetically, as a Border collie, Yasha was an incredibly quick learner. By his eighth month of life, he had learned all the novice and open class obedience exercises of the American Kennel Club and was working on the utility exercises.

As a German shepherd, Yasha was fearless. One day when he was a little over a year old, I went with members of Wofford College's Adventure Club to canoe a few stretches of rapids on the Green River, only about an hour's drive from the Wofford campus. I had become a keen paddler since we moved to South Carolina, and introduced many students to white-water canoeing and kayaking through the Adventure Club. Running class three, four, and five rapids on rivers in the Southeast was a passion of mine, and kayaking the Grand Canyon one year was a thrilling high point.

On this beautiful spring day, Sally and Debbie came along to run shuttle for us, dropping us off at the put-in point upstream and meeting us at the take-out point downstream. Yasha was along as well, full of excitement. He was getting near his full size, but still very much a puppy in temperament. It was not his first trip to the river, but until now he'd mainly only splashed in shallow pools. What swimming he had experienced was in flat, calm water, and although he seemed to love getting wet, he was a weak swimmer at best. So we weren't expecting him to work on his dog paddle that day.

Along a stretch of the river called Big Corky, the rapids drop fifteen feet over a distance of about a hundred yards, and are rated class three — "Intermediate." Sally and Debbie walked Yasha down to the bank to watch us navigate that section. I was three-fourths of the way through the rapids when Yasha leaped excitedly into the water. He had never seen me paddle through white water before and he was eager to join me.

As soon as the current caught him he knew he was somewhere he didn't want to be, and he paddled with all his might, flailing furiously, trying to get back toward Sally and Debbie. Still a growing puppy, Yasha didn't have the strength to get across the current. A whirling eddy sucked him under. He popped back up, struggling to make headway, clawing at the water.

Coming through the last bit of rapids, I threw my paddle into the center of the canoe, grabbed the gunwales, and vaulted out into the water. I gripped the bow of the canoe with my left hand and with my right hand reached out for Yasha, who was just getting sucked down again. I caught the scruff of his neck, hooked my fingers in his collar, and lifted his head above the water. And then I kicked furiously with all my might, levering down on the bow of the canoe with my left hand to buoy Yasha and me up. A few more kicks and we were in quiet water where I could stand up and Deb waded in to lend a hand.

Yasha shook himself furiously. He was panting hard, but his breathing soon evened out and his dominant emotion seemed to be intense excitement. A few minutes later he was splashing around in the shallows, diving for rocks. From that time on, whenever Yasha saw me put the canoe or kayak on the roof of the car, he was raring to go. We didn't want to take chances with him, so Robin sewed a life vest, packed with pieces of flotation foam, that we made him wear whenever he came on the river with me.

Throughout his life, Yasha exulted in clambering rocky riverbanks, nails scratching on the wet rock, and in going swimming with me. It wasn't all that long after his misadventure at Big Corky that he was impatiently watching several students and me body surf the Chattooga River at the bottom of a rapid called Bull Sluice, where the waves were harmless and void of any danger. I was just jumping off when a student hollered, "Dr. Pilley! Dr. Pilley!" I hit the water and spun myself around in time to witness Yasha jumping off the six-foot boulder after me. Thus was born his unquenchable thirst for body surfing.

Only one other activity appealed to him as powerfully — Frisbee play. Bring out a ball and he might simply lie down. Bring out a Frisbee and he erupted into excited barks, jumps, and tornado-fast spins. Whether the Frisbee was thrown directly to him or to his side, Yasha always leaped forward in anticipation of it — the mark of an elite athlete. For the first four or five catches, his return for another throw was rapid. However, after the fifth or sixth throw, mouthing the Frisbee triggered his chewing instinct. No amount of soft or hard recalling could overpower this behavior, an example of what animal scientists call instinctive drift.

Instinctive drift must not be punished. The best way to inhibit an undesirable instinct is to trigger a competing, more powerful instinct. Fortunately I learned that sailing another Frisbee to Yasha made him drop the first Frisbee in order to catch the second one. I also learned that if I ran toward him, he cleverly nested the two Frisbees together upside down and then ran away with both of them in his mouth. I continued this experiment with more Frisbees and discovered that

Yasha could nest as many as six Frisbees together and hold them in his mouth while playing keep-away. Of course, my pace of running was quite slow, to give him time to nest the Frisbees together, but fast enough to motivate his possessive instinct.

Yasha's enormous energy, boundless curiosity, and quick learning — not unique to Border collies, of course, but so typical of them — made me think he would be the perfect subject for animal learning experiments with my students. Until then, my research and teaching as a psychology professor at Wofford College had involved rats and pigeons in Skinner boxes, named for B. F. Skinner, the influential behavioral psychologist. A Skinner box is an enclosure in which the animal learns to press a lever or perform some other behavior to get a food reward.

The experiments produced interesting, statistically significant data, but nothing earth shattering. Never what I was always looking for: an aspect or principle of animal learning that could be generalized to all learning. That was my quest after I left the Presbyterian ministry and found my true calling as a research psychologist and a college teacher. Entering graduate school in psychology at the age of thirty-six, with two young daughters, was only possible for me because of Sally's belief that I should follow my passion to understand more about learning. We were "as poor as Job's turkey," to quote Sally, and for a few years her work as a nurse was our main source of income. But she never complained, and those were happy times for all of us.

In addition to their role in my research, the rats and pigeons in my lab were subjects in countless student experiments. Yasha's rapid learning, in tandem with the long-evolved social bond between dogs and people, offered me an alternative. I was fascinated by the canine intelligence that made the dog-human bond possible and led to the astonishing variety of roles that dogs take on as working and service animals and pets. Border collies' capacity for learning, which I observed in herding trials and demonstrations, particularly impressed me, and I hoped that animal learning studies with a Border collie would be more likely to produce data that could be generalized to all learning. Even more important, I figured that my students and I could have a lot more fun with a Border collie than with rats and pigeons.

Yasha exceeded my wildest expectations. He quickly became not just a subject for student experiments but my full-fledged teaching assistant. If my students and I didn't discover something new about learning with Yasha, it wasn't because of any lack of capacity on his part.

Every fall I taught a course on scientific methods in psychology for non–science majors. As someone who came to science late after a first career in the ministry, I treated the course, an introduction to the principles of learning for both humans and animals, as an opportunity to inspire students to consider devoting themselves to scientific discovery. When it came to showing college students that science was an interesting place to be, Yasha was my secret weapon.

During the first class of each semester, I entered the classroom with Yasha at my heels. As I dropped my briefcase on the desk at the front of the room, Yasha went around and introduced himself, tail wagging. His manner was less solicitous than it was bold and self-possessed, but as always I noticed the extra time and encouragement he gave shyer students, bowing his front legs, angling his body sideways, and tilting his head as he looked up at them in order to emphasize his desire to make friends.

Having established a positive connection with each and every student, Yasha looked at me. I nodded with approval, and he climbed up and sat in the chair closest to the door. As the last-minute arrivals came in, Yasha climbed down to greet them. No student could pass without a successful let's-get-acquainted moment, and I marveled as always at Yasha's confident social intelligence.

The bell in the tower of Main Building, familiarly known as Old Main, tolled eight a.m. I gave Yasha another nod. He sprang down from his seat, with its half-desk writing surface, to paw and nose the door shut. And then he climbed back up in his place and sat as if he were ready to start taking notes.

"Hello," I said. "I am Dr. Pilley, and the dog who greeted you is Yasha. He is half Border collie and half German shepherd. For anyone who doesn't know about Yasha's role in the course"—this elicited smiles and chuckles from the students, most if not all of whom had been attracted to the course, despite my relative stinginess with A's, precisely because of what they'd heard about him—"he serves as my teaching assistant. If you graduate from working with human subjects in the first part of the course, Yasha will also serve as the subject for your individual and group research projects in the principles of learning. Whether he is a cooperative or uncooperative subject will depend on the relationships you build with him. Until then he will have other tasks in the class."

I paused, holding the students' attention, and then turned to Yasha and said, "Yasha, if any students fall asleep in this class, nip them in the ankles."

He barked and nodded his head in reply, and the students giggled, a few of them a little nervously. I knew they'd probably heard some wild stories about Yasha and were wondering which ones to believe.

In the fall semester I taught second-year psychology majors experimental methods employing both animals (rats and pigeons) and humans as subjects for experiments. In the spring I taught my favorite course, devoted to human and animal learning processes such as classical and operant conditioning. After much thought, I decided that I would continue to use rats, pigeons, and humans as the primary research subjects for my upper-level psychology courses — leaving the door open for some of the majors to work with dogs. So it was that Yasha officially became my research assistant.

One course was set up so that students worked in groups of three or four to try to teach Yasha new behaviors, and then demonstrated the results to the rest of the class. This was so successful that I soon expanded the scope a bit by letting students use their own dogs, if they had them, as a few students who lived off-campus generally did. A couple of years later Sally got a purebred female German shepherd we named Grindle, who also became an experimental subject until arthritis made her infirm.

To teach Yasha, Grindle, or their own dog a new behavior, the students had to employ classical conditioning and operant conditioning, the basics of learning for all creatures. In a nutshell, classical conditioning involves creating an association between two stimuli in order to elicit an involuntary response from an animal. Ivan Pavlov made the Pavlovian response famous by pairing the sound of a bell with the sight and smell of food, conditioning dogs to salivate in anticipation of a meal whenever they heard the bell. In operant conditioning an animal learns to associate a given voluntary behavior with a given consequence. Operational conditioning is at work when a rat or a pigeon in a Skinner box learns to press a lever to get food.

Before my students graduated to working with dogs, they first had to do classical and operant conditioning experiments with people. This could mean conditioning a roommate to wear a particular shirt for days on end by saying how good it looked, conditioning another professor to end class before the bell rang by organizing other students to close their books five minutes early, or evoking some other behavior with a cue that the person involved didn't recognize consciously.

For example, students working in teams of two used positive reinforcement to motivate each other to give at least five compliments for each of the next five days to people on campus. In their written lab reports the students had to identify the ABCs of each compliment: the *antecedent* situation preceding the compliment, the *behavior* that was complimented, and the *consequence* of the compliment for both the giver and the receiver, as well as the specific reinforcement that each student used to motivate his or her partner to give the compliment. These ABCs are a good tool for analyzing the two ways to influence any behavior: altering the antecedent situation that triggers the behavior or the consequences that follow it.

The first time I introduced this assignment, I asked the students to compliment the person sitting next to them. I then asked the class to share what they had learned from that. There was silence until a female student said, "I learned that the guys have a lot to learn about giving compliments." That sparked laughter and several valuable comments, mostly by the women, such as that a compliment should be sincere, embody emotion, be something that you think is important to the other person, and refer to something that the person has done as opposed to an aspect of their personality or appearance. The last idea evoked a heated debate, always a good sign that a class is engaged.

Students were often initially flummoxed about how to work with Yasha and Grindle. "Gee, Dr. Pilley," they said, "Yasha and Grindle already know how to sit, lie down, stay, come, and all of that on cue."

"So think of a twist," I said. "Think of getting them to do something they already know on a unique signal. Or think of something entirely new. This is your chance to be creative."

If students continued to have trouble starting to train, I steered them to Grindle and coached them in training her in a variation of a basic obedience behavior, such as sitting in response to another word besides "sit." Grindle was much more patient than Yasha, who adopted a piercing, high-pitched bark when he wanted students to speed things up so he could earn a treat.

Training a single behavior was just a prelude to the major training assignment, which was to teach one of the dogs to complete a chain of heterogeneous behaviors on a single command. The idea was to string together behaviors that a dog was very unlikely to emit in sequence naturally, such as walking a precise route and stopping to sit or lie down at designated locations, and put them all under stimulus control. Grindle and Yasha both learned impressively long behavior chains in this way. I don't think Yasha ever forgot anything students taught him, but Grindle's working drive and memory didn't match his. Without continued repetition, she forgot much of what students had taught her the previous semester, especially over the summer break.

One thing my students taught Grindle that she unfortunately never forgot was how to open doors with round doorknobs. Brass doorknobs throughout the Psychology Department were soon dented with Grindle's tooth marks. The Maintenance Department had to replace the knob on my office door numerous times, owing to a big, powerful German shepherd twisting it around. Occasionally Grindle managed to click the catch on a door in such a way that she locked herself inside. A pair of brand-new boots, handmade in Spain, met an untimely end when Grindle locked herself in Debbie's room and had nothing to occupy her but chewing on some beautiful leather.

Grindle taught herself to open the unlocked door of our Volkswagen Bug from the outside. No doubt she was inspired to do this by her success with round doorknobs. But it took cleverness for her to figure out how to push the pushbutton outside the handle of one of the Bug's doors in order to pop it open. This caused us a lot of worry, because

Grindle couldn't open the door from the inside and could easily get trapped inside on a hot South Carolina day.

There were occasionally other problematic results of the students' interactions with Yasha and Grindle. For example, Yasha usually wandered around a little at the beginning of a class, unobtrusively soliciting treats from students who were working with him in my lab, and then settled down quietly in the back of the room. However, one day he interrupted my lecture three times with his unignorable high-pitched bark. Three times I corrected him. A few moments after the third "Hush, Yasha," as I paused in my lecture, everyone heard a male student whisper, "Frisbee, Yasha. Frisbee." Students will be students, and we all had a good laugh.

There was also the time that one of my favorite students, Gervais Hollowell, and two of his classmates taught Grindle to answer the telephone in my office. I missed quite a few calls until Gervais and I solved the problem by putting the telephone on a cabinet where Grindle couldn't reach it. However, Grindle was also picking up a ringing telephone at home. When I was somewhere in the house or out in the yard where I couldn't get to the phone when it rang, I heard the ringing stop after three or four rings and assumed Sally had answered it. Meanwhile, she was often assuming the same thing. It could be hours before we found the phone off the hook. After we put it back, it inevitably started ringing with calls from friends or family members who had been trying to reach us and getting a permanent busy signal. It took a few weeks of not being rewarded for picking up a ringing telephone before Grindle began to forget about doing it.

Yasha really had a remarkable memory. One year a student took him to the opposite side of the Wofford campus from my office and lab and put a Frisbee in a tree. A month later the student asked Yasha to find the Frisbee, and he took off like a shot straight for that tree.

Over the years, Yasha also learned to pretend to jump over an invisible hurdle, balance a book on his back while walking, climb a ladder, and obey commands delivered by walkie-talkie, among numerous other behaviors.

Yasha was so quick to learn new behaviors and solve new challenges that I wondered if I could teach him to understand words, specifically the names of objects. My procedure was to place two objects, such as a rope and a ball, on the floor about two feet apart. Standing a few feet from the objects, I randomly showed Yasha a rope or a ball identical to the ones on the floor and asked him to fetch the corresponding object. I reasoned that seeing the rope or ball in my hand and hearing me name it would enable Yasha to quickly associate the object and name and retrieve the correct object.

Boy, was I wrong. After a hundred trials, Yasha failed to show any sign that he was learning to retrieve the objects by name. I modified our procedures over a dozen times, but still failed to produce positive results.

My efforts with Yasha never got anywhere, but I always suspected that this was because my method was wrong. I believe Chaser's success in language learning indicates this, because as bright as she is, I don't think she is a whit smarter than Yasha was.

In any case, Yasha continued to be a great teaching assistant and experimental subject for students. He had the freedom of the Wofford campus, although this occasionally raised a few eyebrows. One day a senior member of the campus security detail came to my office as I sat grading exams. This man was fondly known as Deputy Dawg for his success in solving campus thefts. He usually gave me a big smile whenever he saw me, but on this day he looked serious.

"Where's Yasha, Dr. Pilley?" Deputy Dawg asked.

"I don't rightly know at the moment," I said. "He was here not too long ago, but then he wandered off. Is there a problem?"

"Don't you think it could be a problem having him wander everywhere he pleases, looking for mischief?"

"Well, I . . ."

A smile broke out on Deputy Dawg's face. "I'm just funnin' with you, Dr. Pilley. I remember when you first brought Yasha on campus as a puppy and he palled around with the dean's old hound, who already had the run of the place. The truth is, I think it's time we recognized Yasha's presence officially, and I've brought you his faculty ID badge."

Deputy Dawg handed me a faculty ID card in a plastic case attached to a lanyard. It read, "Wofford College Faculty/Yasha/Teaching Assistant."

Deputy Dawg said, "I don't expect Yasha to wear that, mind you. But you might want to keep it in a safe place. You never know when there might be a changing of the guard around here, and you'll need to show some whippersnapper that Yasha's official status precedes his."

We laughed over that, and I thanked him. Just then Yasha trotted in and, tail wagging, went up to Deputy Dawg for a pet.

Deputy Dawg said, "I knew I'd be seeing you sooner or later, boy. Everybody knows that if Dr. Pilley is around, you're bound to be close by."

It was true. If I drove to the grocery store for a gallon of milk, Yasha was in the front passenger seat. Unless Sally, Robin, Debbie, or another person was also along, in which case Yasha happily sat in the back.

In hindsight, the match-to-sample method was not the best starting point for teaching Yasha the names of objects. Moreover, I failed to incorporate Yasha's language training into our playful interactions as I had all of his prior training. I certainly wasn't taking good advantage of his social intelligence. That had to be a critical mistake, because language is inherently a social activity.

In 1990, ten-year-old Grindle became seriously infirm from arthritis, and she passed away that year. Sally took it especially hard, but the whole family mourned the loss of Grindle's big heart and calm, soothing presence.

A friend of mine says, "If you get a pet, eventually you get a broken heart." The relatively short life spans of our pets cause us a lot of grief, but they also ground us in a natural cycle of life and death and, if we can accept it, renewal. The spirit of our relationship with one pet lives on in and shapes the spirit of our relationship with another pet, even years later.

I thought I knew that after our family's experiences with our first dog, Fluffy, a collie Sally rescued when she was pregnant with Robin, and then with Bimbo and Grindle. But I still had a lot to learn about loss and renewal.

Border collies often work sheep until they are fourteen or even fifteen years old. Yasha was so vigorous for so long that I found it hard to accept when he began to slow down dramatically. Through the fall semester of 1993 he faithfully and effectively served as teaching assistant and research subject. But at the beginning of the spring semester, I saw that Yasha no longer had the energy to go to Wofford and be at my side day in and day out. I began leaving him home, and our reunions every day were bittersweet.

One cold day in March I sensed something was wrong as soon as I came in the door. "Yasha," I called, and then, raising my voice, because at sixteen years of age Yasha was a little deaf, "Yasha! I'm home."

The low whimpers that came back cut through me. I followed the sounds and found Yasha lying on the rug in our first-floor bedroom. "I've left you alone too long, haven't I, boy?" I said as I knelt down beside him and stroked his head. Sally was in New York City visiting Debbie, and no one else was in the house.

Yasha relaxed under my touch. Looking into his trusting brown eyes and seeing his indomitable spirit flashing there, I asked myself if it was time to let him go. I couldn't let myself think that it was.

"C'mon, Yasha. Let's get you some water, and some food, if you want it." I got to my feet and half turned to the door. Yasha strained to get to his feet, but he couldn't. When I'd left the house a few hours earlier, he'd still been able to get up and walk around.

When I bent down to pick Yasha up, feeling how thin he was gave me another wrench. I carried him into the kitchen, then knelt down with him cradled in my arms so that he could drink a few sips of water. I carried him to the living room couch and sat down beside him, scratching behind his ears. He lifted his head and looked at me with utter trust, then lowered his head to his paws and fell into a light doze. While he slept, I convinced myself it was not yet time to let him go. When he stirred awake intermittently and I saw him become aware of his discomfort, my heart sank. Around ten o'clock, Yasha woke up and started panting. He was suffering, and I knew I couldn't let that continue any longer. I called the emergency animal clinic to make sure a vet was on duty, then I wrapped Yasha in a blue blanket against the cold and carried him out to my pickup truck. It was only a ten-minute drive, and every second of the way I wished it was longer.

The vet examined Yasha and agreed with me that it was time to help him on his way. He gave Yasha a sedative and then an injection to stop his heart. It was almost more than I could stand to see the light go out of Yasha's eyes. Back home, I carried Yasha, still wrapped in the blue blanket from shoulder to tail, into the bedroom and put him on top of the bed. Fully clothed with my shoes still on, I lay down beside him and turned out the light.

The next morning I went slowly about the house in a daze, picking up and putting down Yasha's toys: various Frisbees, all of them chewed around the edges, especially his favorite yellow one, and a stuffed rabbit with one eye and torn ears that Sally had mended over and over again. I was in no hurry for what came next. But it had to be done. I got a shovel and went out behind the house to where Grindle was buried. It was a grassy spot shaded by three oaks and a pine. Turning back to face the house, which I'd built myself in a log cabin style, I remembered how Yasha had followed me up the ladder one day when I was finishing the roof and I'd had to carry him back down.

I started to dig. The effort released the tears I'd been holding back for hours, and I had to stop several times and lean on the shovel to prop myself up. Finally the two-by-three-foot grave was ready. I went inside and got Yasha, still wrapped in the blue blanket, and gently laid him beside the grave. I went back inside for his favorite yellow Frisbee, the stuffed bunny, and his brown leather collar, which he hadn't been wearing because he was so thin.

I sat down beside Yasha and his things, covered my face with my hands, and wept. When my tears finally stopped and I looked up, the sky was lowering and threatening to pour with rain. I couldn't wait any longer. I picked up Yasha and embraced him one last time, then laid him in the grave with his yellow Frisbee, stuffed bunny, and collar. Slowly, I picked up the shovel and began to fill the grave with dirt.

I knew friends would tell me to get another dog, another Border collie, soon. And I knew I couldn't do that. I told myself that I would never get another dog, and I allowed myself one more cry.

Walking slowly toward the house, I recalled my grandmother's words whenever we came home from a funeral service for someone near and dear to her. As the front screen door slammed behind me I heard her holler, "Child, it's not the letting go that hurts so much, it's the holding on!"

I wasn't ready to let go. So I decided to gingerly lock this pain into a black box inside me, where it would be safe until I could fully resolve it.

I went inside to call Sally and tell her what had happened.

3

Christmas in June

IN MY HEART I knew that we needed to get another puppy and renew the life cycle of raising a dog as part of our family. But after sixteen years with Yasha almost always at my side, a voice in me said, "Hold off. Give yourself a little more time to enjoy your memories of Yasha."

For now at any rate, I had to move forward. Spring semester was under way and I had classes to teach. And Sally and I still had dogs to take care of and love. Robin had acquired two purebred Siberian huskies as puppies. The first of them was a big handsome male she named Blue, for the hint of blue at the tips of his gleaming white fur. The second, a slightly smaller, no less handsome female, was Timber, named as much for how the sound of the word fit her easygoing disposition as for the tree-trunk gray in her fur.

Robin's primary occupation was as a head white-water rafting guide at the Nantahala Outdoor Center in western North Carolina. Nantahala is the largest white-water outpost in the eastern United States, and Robin's position was active for eight months of the year. During the winter months, Robin worked in New York as a photographer and lived in Brooklyn with Debbie and her husband, Jay. Because Jay is allergic to dogs, Blue and Timber came to live with Sally and me during these months. There were also times when Robin was working on the river many days in a row, or guiding trips on rivers in Tennessee and Georgia, and we were happy to babysit the dogs then too.

Blue and Timber were fabulous escape artists. Blue especially was an alert, inventive problem solver in that regard, whether it meant nudging a gate open or finding the best spot to dig under a fence. He got out of the yard, with Timber happily following his lead, on numerous occasions.

Although we generally discovered they were missing within minutes, there were many routes they could take through the neighborhood, and they ran like bullets. We were very fortunate that most strangers found the duo charming, and we usually tracked them down without the authorities getting involved. We knew that if we found Blue, Timber would be at his side. And the best place to find Blue was at a playground making friends with children. But on two occasions they were caught by the dogcatcher far on the other side of Spartanburg. We were enormously relieved to get the dogs back, of course, but we didn't like the fines, which doubled with each offense.

Blue and Timber became experimental subjects for students as Yasha and Grindle had before them. In coaching students to devise and carry out experiments with Blue and Timber, I emphasized that they needed to work with the dogs' defining characteristics as Siberian huskies and as individuals. As sled dogs, Siberian huskies have been bred to accentuate an unrelenting drive to run and chase down the miles even in the harshest winters. As individuals, Blue and Timber found food treats powerfully motivating. Of course, all dogs love treats, and they're often essential in training, especially in the early stages of shaping and reinforcing a behavior. But for some dogs the innate drive to herd or hunt can far outweigh food as motivation to practice and perform a behavior. However, Blue and Timber, like Yasha and Grindle, never found food treats boring.

Blue and Timber were the most laid-back dogs we ever had. They were also the most capable of pretending that they did not hear or understand a command. To some degree all dogs are liable to ignore commands they aren't ready to follow. Border collies who have been called off the livestock may be so determined to continue herding that the shepherd or trainer has to repeat the command several times. Blue and Timber were masters of pretended ignorance, however. If a student uttered a command that another student had recently taught them, they often acted as though they were deaf. Fortunately a yummy treat or a friendly but firm shake of neck fur instantly restored their hearing. Despite their escape behaviors, Blue and Timber were gentle, good-natured dogs, and students always enjoyed working with them.

Students came up with ingenious ways to shape the behavior of Blue and Timber, working with each dog separately. For example, two different groups of students trained Blue and Timber, respectively, to pull a light cart in a smooth sequence of gaits from a standing start to a slow walk, a fast walk, a trot, and a run, and then back through the gaits to a full stop. Another memorable experiment with Blue and Timber involved teaching them a complicated series of maneuvers — another of those chains of heterogeneous behaviors that I discussed in the last chapter — to reach some food on a high cabinet in my lab. To get the food, each of the dogs had to pick up a rope tied to a chair in its mouth, pull the chair into position next to the cabinet, climb up on the chair, and then stand on hind legs with front paws stretching up onto the front of the cabinet.

Given their experience with human subjects, students could see that they were using the same principles of reinforcement with the dogs. The learning might consist of very different responses of very different complexity. But reinforcement of a behavior increases the probability of the behavior. And nonreinforcement of a behavior, like not rewarding Grindle for picking up a ringing phone, decreases the probability of the behavior. In terms of the ABCs of behavior I mentioned earlier, consistent nonreinforcement motivates a long-term change in behavior by altering the consequences of the behavior rather than the antecedent situation.

I stress long-term change, because if a dog has been begging successfully at the dinner table, for example, and then people stop feeding the dog from the table, the dog will likely beg even more in the short term to try to get his or her message across. It's just like what happens when parents who have been providing candy bars in response to a child's tantrums in the grocery store start ignoring the tantrums. The child will likely throw the worst tantrums ever soon after that. Inexperienced trainers or parents often abandon nonreinforcement then, either by caving in and providing the treats or by resorting to punishment. Both approaches are doomed to fail.

However, the begging-food-from-the-table dog and the throwing-a-tantrum-to-get-a-candy-bar child both will respond to consistent nonreinforcement of those behaviors by experimenting with other behaviors, and some of the new behaviors will be positive and desirable. Amazing learning results when trainers, parents, and teachers reinforce such positive, desirable, freely chosen behaviors, because this motivates the learners to experiment with other new behaviors that might win reinforcement. I'll come back to this global principle of learning later in the book, because of how it animates creative learning.

I was sixty-six when Yasha died, but I put off retirement for two more years because I didn't know what I wanted to do next. Throughout my life I'd made a point of trying something new every five to ten years, whether it was a shift in my research and teaching or taking up new sports such as kayaking and windsurfing.

In my last several semesters of teaching, I changed things around as much as possible, seeking new perspectives for the students and myself. I even began holding discussion sessions in volunteers' dorm rooms. Cramming fifteen to twenty students together in a small dorm room made for intense conversation and offered wonderful opportunities to get to know each other better. The casual nature of the sessions encouraged even the shiest students to voice their opinions and challenge their peers.

I enjoyed breaking out of the confines of the classroom and traditional class structures, but I also remained restless to find a new focus for myself after I retired.

Somehow, putting off getting a puppy became part of that. I kept thinking I needed to find something entirely new to do, something I'd never done before and could become passionate in pursuing.

In the summer of 1996, at the age of sixty-eight, I finally launched myself into the abyss of retirement. I figured that once I was retired, something would jump up to grab my interest. Nothing did, or at least nothing that really excited me intellectually. I stepped up my windsurfing excursions to Charleston. And I became a regular up at Robin's house at the Nantahala, kayaking with her during her downtime.

I also stayed abreast of what was happening in animal behavior research. I continued to be fascinated by the intelligence and learning potential of dogs in general and Border collies in particular, especially with regard to possible language learning.

Because of the complex commands in herding livestock, herding dogs seemed the best possible canine candidates for language learning experiments. And Border collies struck me as the pinnacle of herding dogs.

My friend Wayne West did a lot to deepen my understanding of Border collies. Wayne was at that time a firefighter in Spartanburg, and he raised sheep and bred Border collies on a small ranch, Flint Hill Farm, that has been in his family since colonial days. A tall, burly man as comfortable on a cutting horse as most people are in their easy chairs, Wayne has a somewhat high-pitched twang and an infectious, rather dry sense of humor. Over the years I often took students on visits to Wayne's farm, or to the Border collie exhibitions he organized every two or three years at the Spartanburg County Fair in October. I never missed one of Wayne's exhibitions, which he also sometimes mounted in a smaller way on his farm.

Wayne taught me the fundamentals of herding. Farmers and ranchers like Wayne have been breeding Border collies for centuries, selecting dogs for their ability to keep their eyes on the sheep and listen to the farmers. If a farmer's dog didn't listen to the farmer, the farmer didn't breed the dog, and gradually Border collies gave greater and greater attention to the words of the farmer. Through one generation after another, the herding instinct was shaped in Border collies just as hunting instincts were shaped in hunting dogs. When Border collies are first introduced to a flock of sheep, by instinct they will move out to the opposite side of the sheep and begin trying to herd the sheep to the farmer without any training.

There are only a few herding commands that the dog needs to learn. "Go out" means take a wide arc out toward the sheep. "Come by" means go clockwise around to the other side of the sheep. "Way to me" means go counterclockwise. "There" means freeze instantly in a standing position. "Drop" means drop into a crouched position that enables immediate action. "Walk on" means walk toward the sheep to get them to move toward the farmer. For most situations the stern "eye" of the Border collie is sufficient to move the sheep. In those situations where a sheep refuses to be herded, a quick nip on the hind leg by the dog will elicit immediate cooperation.

In 2000 Wayne got several Border collie trainers to do an exhibition and judge herding trials at the county fair. I spent all day there, captivated by the joyful intensity of the dogs in their work and the closeness of their bonds with their trainers. That evening I had the privilege of sitting around the campfire with Wayne and the other trainers, breeders, and sheep farmers — and their dogs.

In the course of that evening I told them my research had never found any indication that dogs could learn the names of objects. In my lab at Wofford my students and I tested this with Yasha, Grindle, Blue, and Timber by asking them to fetch a particular item from one of a few different objects. The results were never better than pure chance.

Speaking to these expert trainers as if they were students in one of my classes, I described how dog owners insisted to me over the years that their dogs understood the independent meanings of "newspaper," "ball," and the names of other objects they could retrieve on command. I didn't see why dogs couldn't learn such things and I hoped these people were right, so I always invited them to bring their dogs into the lab for a test under controlled conditions. But as with my own dogs, the results were always negative. My conclusion was that in daily life, the situation guided dogs as to what to fetch. They were never really faced with a choice of what object to go pick up and bring back.

In telling Wayne and his friends about this, I added that as far as I knew, no other researchers had ever documented anything different. My research seemed to indicate that dogs didn't even understand their own names except as signals to pay attention.

They looked at me as if they thought I was nuts, but out of deference to Wayne they didn't say so. One crusty old farmer said, "So that's what science says, huh?"

Another sheep farmer said, "I'd appreciate it if you and the other scientists that have looked into this could explain something to me. If Rascal here"— he nodded at the mostly black Border collie lying at his feet —"doesn't understand any words or names, how come I can ask him to herd sheep out of the flock by their names and he'll do it without a hitch, and without my gesturing or even looking at the sheep?"

I heard assenting grunts and murmurs, and I saw Wayne look at me with the barest hint of a smile crinkling the corners of his eyes and turning up the edges of his mouth in his broad, good-natured face. The trainers' polite but emphatic reaction humbled me, and rightly so. It was one of a number of experiences, including reading books by Border collie trainers, that convinced me there was much I could learn from people who worked with dogs in traditional ways. It made me think once again that the failure of my experiments in trying to teach Yasha the names of objects probably had more to do with my flawed methods than with his learning abilities.

In 2000, Blue, then fourteen years old, developed cancer. Robin laid him to rest in the backyard near Yasha and Grindle.

Timber was really lost for a time without Blue. But she transferred her love and devotion to Sally. Every dog we had loved Sally, including Blue and Timber. But now Timber followed Sally throughout the house obsessively, as she had previously followed Blue. "Don't you get tired of having Timber under your feet?" I sometimes teased. Sally never shot back, as she might well have done, "Did you get tired of having Yasha under yours?"

Late in 2002, Timber, then eleven years old, also fell ill with cancer. Robin and Sally and I mourned Timber's passing together. I was proud of how Robin had raised both Blue and Timber up to be such loving dogs.

With Timber gone, Sally and I were without a dog in our house for the first time in more than forty-seven years — since we were newlyweds in 1955. Despite the void this created, neither Sally nor I suggested getting another dog. But soon enough, I found myself continually bringing up Border collies in our conversations. I marveled at their seeming ability to learn language. I found myself reminiscing about Yasha's learning capabilities, wondering what more he could have picked up with the right training. But I still didn't think of getting another dog, Border collie or otherwise.

Sally has grounded me all our life together, and she knew what she and I needed even if I didn't. At Christmas in 2003 we were sitting around the tree, watching Aidan play with his new Thomas the Tank Engine trains. Deb was sorting out the pile of empty boxes, and Robin was organizing the rest of the cleanup.

I said I was going to try out my new tool belt. Sally stood up and placed herself in my path. Hands on her hips and with a crooked smile on her face, she announced, "John Pilley, you're getting a Border collie puppy for Christmas. But you have to wait until sometime this spring before we can go out to Wayne West's and bring the puppy home. He or she hasn't been born yet. I don't want to hear any arguments. This is already done."

If she had told me that grass was blue and the sky green, I couldn't have been more surprised — or delighted. A puppy, a Border collie puppy, and one of Wayne West's Border collie puppies. This was a merry Christmas!

Sally mistook my blank look and stunned silence for resistance. Her southern drawl became more pronounced as she quickly explained, "I called Wayne a few weeks back and told him I wanted to get you a puppy." One of several reasons Wayne is renowned for his Border collies is that he is a highly selective breeder. He told Sally he wouldn't have a litter on his farm until the spring, and offered to refer her to another breeder, an excellent one. Sally, God bless her, said, "No, Wayne, it has to be one of yours." As I slowly stood up, she raised her chin. Looking me straight in the eye, she said, "And that's all there is to it."

In a second I had her wrapped in my arms. "Thank you, babe," I whispered in her ear. When I finally let go of the hug, we were both grinning from ear to ear. She grabbed my hand and tugged me to the kitchen, saying, "Come on, that turkey isn't gonna carve itself."

As winter passed into spring, I grew more and more excited about getting a new Border collie puppy. I wondered how far a Border collie could go in learning the meaning of words and other aspects of human language if I poured the training on in puppyhood. What methods could best trigger and extend such learning?

I had no idea, really. There was nothing in the scientific literature that said it was possible or gave a hint as to the best approach. I decided that I would have to take my cues from the dog, and try to devise a method that suited the breed-related and individual attributes that he or she displayed.

At the end of April, Wayne called and said, "We've got a litter of seven pups. You should have some good 'uns to look at in a month or so, when they're ready to have visitors."

By the end of May we couldn't wait any longer and drove out to Wayne's farm to look at the puppies. It felt like Christmas morning all over again as we approached the mass of wiggling black and white fur balls at the side of their mother, Tess, who was curled around them in a protective U shape. Tess shot us a quick guarded look as she licked the face of a rubber-legged pup with a black patch on one eye. Her babies were still too young to be separated from her, their eyes barely open as they stumbled over each other, searching for her milk. Sally and I looked on in admiration and awe, making no attempt to steal a cuddle or choose among the pups. It wasn't until we were in the car that we broke our silence.

"Well, in another month," I began. Sally finished my sentence with a giggle: "We'll have a puppy!" We didn't stop grinning the entire ride home.

Slow as Christmas coming when you're a little kid, June 28 finally arrived, a beautiful, sunny day. Now eight weeks old, the puppies were weaned off their mother's milk and ready to leave her side. At Wayne's farm, Sally and I sat under a big oak tree and Wayne brought the puppies and their mother out onto the grass about ten feet away from us. He was offering us the pick of the entire litter.

We watched the puppies toddle on all four legs while they sniffed in the grass and wriggled around each other and their mother. Then one of them, mostly white with black markings, including a black patch around her left eye, looked over at us. The puppy came over, not just tail but whole hind end wagging, and climbed into Sally's lap. We now saw that the puppy was a female, and she sat happily in Sally's lap while we petted her for a few minutes. And then she made her way back over to her mother and siblings.

Watching the puppy in Sally's lap, I felt the spark of magic as her little brown eyes looked up at me. Most animals are drawn to Sally, but this little one was singling us both out. A connection had been made, and for whatever reason, this puppy had chosen us.

I felt almost giddy as I said, "So this is the one?" It was really more a statement than a question. Sally took my hand and smiled. "Yes!"

When we drove away from Wayne's farm, we took the newest member of our family with us. Wayne cautioned, "These pups are apt to get car sick, you know." But as Sally says, the pup sat in her crate on the back seat "as pretty as you please." Of course, Sally sat in the back seat next to the pup, softly talking to her until she curled up, her little eyes growing heavy as, rocked gently by the motion of the car, she surrendered to sleep.

As soon as we got in the house, Sally and I knelt down beside the crate on the living room floor. We opened the crate door, encouraging the puppy, "Come on! Come on, baby!" Without a shred of hesitation out walked our beauty, and we playfully began introducing her to her new world.

The puppy's arrival showed us how big the void in our lives had been over the past few years. Now there was a fresh young spirit in the house, a creature eager to discover the world and her place in it. The puppy's enthusiasm for each new day fired up the same enthusiasm in both Sally and me. But especially me. I told Sally, "This is the best Christmas present ever!"

We couldn't wait to see how her personality developed day by day.

4

Let the Puppy Be a Puppy

I WOKE UP THINKING about the puppy and the Lobster King. The story of the Lobster King is a parable told by Bernie Dunlap, a former Rhodes Scholar who came to Wofford in 1993 as a chaired professor in the humanities and became president of the college in 2000. I sat in on all of his classes after I graduated from teaching in 1996. One day he began his class by saying, "When I was in college, the Lobster King came to me in a dream and said, 'Bernie, you have been good. I am going to grant you three wishes. The first wish is you can have great wealth and power.'"

There was a crowd of people in Bernie's dream, and they all cheered that wish. Bernie's dream self said, "Thanks, but that doesn't interest me."

The Lobster King said, "The second wish is you can live forever."

Again the crowd cheered. Bernie said, "Well, forever is a long time. I will need time and experience to think about that."

The Lobster King said, "The third and last wish you can choose is to have eloquence. When you speak, the sun will shine, flowers will bloom, birds will sing, and the music of the heavens will play."

The crowd booed. "What kind of wish is that?" they yelled. But Bernie said, "I like it!"

With his many talents, Bernie could have amassed personal wealth and power. But he used his truly remarkable eloquence to become an inspiring teacher, and as president of Wofford he has inspired the entire college community.

As Bernie related the dream and discussed its implications, an hour's lecture seemed to pass in a few brief minutes. The whole class loved Bernie's story, which is really about aligning our ambitions with our aptitudes — getting in touch with our unique genetic inheritance. In several lectures that I have been invited to give at Wofford since my retirement, I have recounted Bernie's dream to students and asked them, "What three wishes did the Lobster King give you?" Sad to say, usually half the students say they have no idea what their own best aptitudes and ambitions might be. Too often, parents, teachers, and the culture, with their needs and demands, mask or dim students' awareness of their own talents and interests. I always hope the story of the Lobster King can be a vehicle for challenging students to discover or rediscover their gifts.

The Lobster King, whom I naturally pictured as an enormous lobster wearing a jeweled crown, had been on my mind since the day before, when we brought the puppy home from Wayne West's farm. What special gifts did she have, and what was the best way to develop them? Did her gifts include an ability to learn the meaning of words?

I knew without a doubt that as a Border collie, she had genes for herding. Centuries of breeding by farmers had endowed her with the capacity to keep her eye on the sheep while giving her ear to the farmer, and I felt that by rights the Scottish mountains should be her home. At the same time, I felt strongly that she had chosen Sally and me as her guardians. I silently promised her that Sally and I would let her be herself, and that we would do our best to make her life a happy one.

I fumbled for my glasses and checked the time: five a.m. That was actually a little late for me. I've been getting up before dawn most of my life. But I'd been restless and had difficulty getting to sleep the night before, with the excitement of having a puppy in the house for the first time in twenty-six years.

I flicked on the lamp by my side of the bed. Sally was still sleeping. She generally went to bed and got up a couple of hours or so after me. And the puppy was still sleeping in her crate.

Sally and I were ambivalent about confining her in the crate during the night. People didn't commonly crate train their dogs when we'd last had a puppy. But Wayne West recommended we give it a try as an aid to housetraining and establishing a good daily schedule for the puppy. "She'll treat it like her own little den and be less likely to have an accident," Wayne told us. The puppy had whined off and on for a few minutes after being put in the crate the night before. But then she had settled down and gone right to sleep on some soft throw rugs, with a couple of toys beside her that were safe to mouth if she wanted.

The pool of light from my reading lamp reached into the front of the puppy's crate. She was sleeping with the right side of her head on her paws, and I could just see the black patch on the left side of her face merging into the shadows. More of Wayne's words came to mind: "All dogs are sensitive to how you speak to them. But the Border collie just might be the most sensitive. The quieter you speak to one of these dogs, the more they'll concentrate on listening to you to figure out what you want them to do. And then when you do raise your voice in praise or correction, it'll mean something."

I loved Wayne's practical wisdom and country eloquence. Before we'd left his farm with the puppy, I'd asked him if he had any special advice for us.

"Only the same special advice you've heard me spout again and again over the years we've known each other, Doc Pilley."

"Go ahead and tell me again. I like to hear you say it."

Wayne grinned and said, "Let the puppy be a puppy. Give the dog time to develop, and piggyback your training on her natural instincts. I tell people that want their dogs to work livestock, 'Take a young dog around the farm with you. When you put out hay and the stock come to it, let the young dog out of the truck. By instinct they'll run around behind the stock, and they'll think they brought them to you. Praise the dog bigtime for that, and you're on your way.'"

"That's what we're gonna do, and we're on our way," I said with a laugh and a wave, both returned by Wayne, as we slowly drove away.

Wayne's friend David Johnson, a Border collie trainer he often sent his own dogs to in order to complete their training, spoke in the same terms about the value of a quiet, calm voice, using the dog's natural instincts as the foundation of training, the power of praise, and the importance of limiting harsh words and negative treatment. Around a campfire after a Border collie exhibition, I had heard David say, "When a dog does something that pleases me, I brag on him and praise him and let him know that every way I can."

All these things harmonized with Sally's and my prior experiences with dogs, as well as with all I knew about creative learning from my experiences as a psychology professor and experimenter. But theories and principles are only as good as their application. I was ready to play with my puppy.

I slipped out of bed and tiptoed over to the crate, and the puppy stirred awake at my approach. "Good morning, Puppy," I whispered softly as I bent down and opened the crate door. The puppy stood up slowly, yawned widely, blinked and stretched, and then trotted eagerly out of the crate to lick my face and hands as I petted and held her.

Sally and I had no idea what we were going to name her, but we'd decided "Puppy" would do for now. We were both keen to find a name that really suited her, after we got to know her personality fully.

Her bedding was bone dry. Wayne's kennel practices built on a mother dog's instinct for keeping the den clean and teaching the young to relieve themselves outside it. The result, Sally and I knew from yesterday, was that the puppy was basically housetrained on arrival. But I didn't want to make her wait too long to get outside. So I shuffled into my clothes, talking softly to her all the while, and she eagerly followed me out onto our screened-in back porch. I put her on a leash, and then we walked down the porch steps to the backyard. It was still dark outside, but there was a glimmer of dawn in the sky. We walked away from the house along the chainlink fence bordering the yard, and I said, "Come on, girl. Do your business!" I spoke with enthusiasm but in a fairly quiet tone.

I repeated "Do your business" a few times, especially once the puppy began to do just that. As soon as she finished I praised her warmly: "Good girl! Good dog!" She promptly wiggled over and leaned against my legs for a pet, apparently pleased with herself for making me so happy.

I took her back into the house, gave her some water, and poured myself a cup of coffee. While I drank my coffee in the kitchen, the puppy wandered into the living room that opens out from the kitchen and dining area. She gently picked up a stuffed animal that we'd played with yesterday, a little lion with a tail that she enjoyed seeing whip back and forth in the air when she shook it. Holding it in her mouth by the neck, she gave it a little shake, and then stopped and looked at me and came over to me with it.

"Good girl," I told her, and gave her a pet. It was wonderful to see her initiating the play routine and engagement with me. Domestic dogs come into the world prepared to do these things. But how you respond is crucial to reinforcing this positive, playful engagement with you and sustaining it throughout a dog's life.

Kneeling down, I gently moved the toy in her mouth while slowly saying, "Out . . . out . . . out." She gradually relaxed her jaw, and the instant she let go of the toy I reinforced her behavior with a vigorous "Good girl!" and another pet. And then I tossed the toy into the middle of the living room floor. The puppy sprang after it, took it in her mouth and gave it a shake, and brought it back to me.

As she performed each action I named it and praised her performance with quiet enthusiasm. "Take toy; good girl," I said, when she spontaneously picked up the toy in her mouth. "Shake toy; good girl," I said, when she spontaneously shook it. "Here; toy to Pop-Pop; good girl!" I said, when she spontaneously brought it to me. The more I associated the sound of a word in her mind with a specific action, the more that word would become a signal for that action. And, I suspected, the better foundation we would have for teaching her the meaning of the word apart from the specific action of the moment.

Still on my knees, I petted her for coming to me with the toy. She looked up with a grin on her face and held eye contact with me for several seconds, a good bit longer than she'd yet done. Building eye contact with your dog is an essential part of training and establishing a loving relationship. But it can be difficult for dogs to learn to do this comfortably — food is usually necessary as a lure and reward — because in the natural environment a direct gaze between dogs often represents a challenge and triggers a fight-or-flight situation. Sally and

I were amazed by how willing the puppy already was to look us in the eye, and I was delighted to see this quick progress.

"Out," I again said softly two or three times while gently moving the toy in her mouth. She released the toy and I petted her and said, "Good girl! Good puppy!" She wiggled with pleasure and interest, eyes bright and wide, ears pricking up at the sound of my words.

I tossed the toy across the floor a few more times for her to capture and bring to me, reinforcing the results with praise and pets. And then I said, "Let's take a walk, Puppy. You've got to explore your new world." I put her on the leash and we went out the front door. It was only five-thirty a.m. and there was no traffic in our quiet residential neighborhood. Although sunrise was a little ways off, the front yard was bathed in soft dawn light.

There was more than a hint of a typical South Carolina summer's day ahead, but that was fine with me. I love warm weather and it's rarely too hot for me. That's not the case with Sally, and we always have to negotiate how we set the air conditioning in the house or the car. With her thick coat, the puppy was probably going to be more of Sally's mind than mine on that issue, I reckoned.

As we walked across the front yard, a squirrel scampered to a tree and the puppy instinctively chased it. In a firm but not harsh tone I said, "No!" I held the leash as the puppy ran to the end of it and brought herself up short, actually knocking herself off her feet. I knelt down and when the puppy got to her feet gently called her to me: "Here, girl. Here, Puppy. Come to Pop-Pop." She came to me with a confused look on her face, tail and ears down, eyes narrowed.

She immediately brightened back up and wagged her tail as I petted her and said, "Good girl. You came to Pop-Pop. Good girl." I continued to pet her and talk to her softly for a few moments: "You have to learn not to chase squirrels and other little animals. But you'll get to chase lots of other things. One of these days you'll even get to chase some sheep, I hope."

We proceeded on our walk. "This is grass, Puppy. Grass. Grass," I said. And then as we left our yard, "Puppy, this is the street. Street. Street." There were no sidewalks in our neighborhood, and if a car came along I planned to take the puppy onto the edge of a neighbor's lawn.

Talking to her frequently in simple words and a quiet, soothing tone would contribute to her development in a couple of ways, if all went well. First, it would definitely help build positive associations for the puppy with the sounds of Sally's voice and mine, and with proximity to us. That was also the heart of the play sessions we'd begun as soon as we'd gotten her home the day before. Second, it might help to prepare her for learning the meaning of words. Although I wasn't yet as well versed in children's language learning as I soon would be, I knew that children whose parents talked to them a lot throughout infancy and toddlerhood tended to be much quicker, more proficient language learners. I was eager to see what effect the same practice could have with the puppy.

We turned right on Seal Street, the block-long street our house sits in the middle of, and turned right again at each intersection, onto Tanglewylde Drive, Foxcross Road, Briarwood Road, and back onto Seal Street again. Like all puppies, our puppy was eager to sniff here and there along the walk. A couple of times she wanted to leave her scent and stopped to urinate. She didn't have much left in her bladder, but on each occasion I repeated "Do your business!" in an encouraging tone several times as she urinated, and praised her warmly with pets and repetitions of "Good girl!" when she finished. I told myself it was probably my imagination, a case of wishful thinking, but it seemed that her ears pricked up a bit on hearing "Do your business!"

The puppy pulled and strained at the leash when she wanted to get to something she smelled or saw. I responded as neutrally as possible to this undesirable behavior. I stood in one place and held the leash firmly, but didn't pull back or yank on it, and I moved along only when the puppy came closer to me and released the tension on the leash herself. When she did this I strode ahead and told her, "Good girl, Puppy! Good dog!" Then I let her explore the spot she'd been trying to reach.

This made progress on the walk slow at times, but I managed not to get exasperated and yank on the leash or speak harshly to her when she pulled. Fighting a dog's behavior in this way is counterproductive. Over time it would create strong negative associations in the puppy's mind with me and my actions and tone of voice, and I was determined to avoid that. Our main goal, as the puppy got to know Sally and me, was to create strong positive associations in the puppy's mind with us, our actions, and our voices, increasing the likelihood that she would come to us and stay by our sides when we asked her to.

We didn't see a single car on the road, but the puppy saw and ran after two more squirrels. I responded to each event the same way as before. I said a firm but not harsh "No!" And then I let the puppy run to the end of the leash, calling and beckoning her over after she knocked herself off her feet, and praising and petting her for coming to me. The puppy's third chase had a little hesitation to it, and as we continued our walk after that she strained at the leash a bit less and released the tension on it herself a bit more quickly.

These were signs that she was already learning an important lesson that would make walks easier and help to keep her safe. Although our neighborhood is relatively sleepy, Briarwood Road can be very busy at times. Sally and I knew several neighbors whose beloved pets were hit and killed by cars when they chased squirrels or other animals into the street. In a couple of cases, large dogs had even yanked their leashes out of their owners' grips and then run fatally into the street. Nothing like that had ever happened to one of our dogs, and we sure didn't want it to happen with our new puppy.

As the puppy and I turned in to our front yard, I glanced across the street. A big gray feral cat with a bushy tail was looking intently at us, especially at the puppy. The tip of the cat's raised tail was twitching back and forth, as if in readiness for a pouncing leap. Judging by its size it was probably a young male from a line of Maine coon cats, among the biggest domestic cats. It was about twelve inches tall and almost two feet long, not counting its tail, and it weighed a good fifteen pounds if it weighed an ounce. Our puppy was only eight weeks old. She was scarcely ten inches high at the shoulder and weighed only a few pounds. She'd be helpless if that big cat caught her alone.

The cat had been roaming the neighborhood for a few weeks, making regular visits at the house of a kindhearted neighbor who put out food and water for it. But the cat was also a busy hunter, as we'd seen him racing across the lawn, critter in mouth. Robin said she'd seen the cat near one end of a drainpipe that ran under the street to the woods behind our house. That made me more uncomfortable.

I moved in front of the puppy and stared the cat down, doing my best to appear threatening. Unimpressed, it gave me a nonchalant look and then padded off in between two houses on its side of the street. Before bringing the puppy home yesterday, I hadn't really given the cat much thought. Now I wanted it gone.

Dismissing the cat from my mind for now, I took the puppy into our fenced-in backyard. I still wanted to be able to stop her from chasing a squirrel if she saw one, so I left the leash on her collar but let her trail it along behind her. She sniffed around the steps to the back porch, then trotted off to the sixty-foot-long chainlink fence that separated our yard from that of the neighbors directly behind us. Here she resumed sniffing with great intensity.

"Yes, girl," I said. "Two dogs live on the other side of the fence. You'll meet them before long, I'm sure. They're a couple of friendly creatures."

As if on cue, the neighbors' back door opened and their dogs came out. On spotting us they both ran over to the fence. A large mixed-breed dog was in the lead—the type of dog Sally calls a Heinz 57, a combination of so many varieties that it's hard to pin any kind of tag on it. A little bulldog came panting behind.

The puppy stood stock-still, and I knelt down behind her so that she was sheltered a bit between my knees, with the fence a foot in front of us. She wasn't cowering and didn't seem distressed, but she showed no inclination to engage with the dogs as they came to a stop at the fence, barking in excitement. "It's okay, girl," I told the puppy, stroking her side. "These are nice dogs." They were the first dogs she was exposed to besides her own kennel mates, and it was vital to socialize her so that she was comfortable around other dogs.

The little bulldog and the big mixed breed sniffed through the fence toward the puppy. She remained motionless. The neighbor dogs bounded back and forth in front of the fence, bowing their legs in play postures. The puppy stayed very still, and looked back at me as if to say she'd had enough of this encounter. But then the neighbor dogs started racing back and forth along the fence, and that excited the puppy to race back and forth with them.

I stood up and let them race. The puppy could barely cover a quarter of the length of the fence in the time the big dog took to run to the end and turn back to run the other way, and even the little bulldog was too fast for her. But the puppy plainly loved to run. Her exhilaration made me smile.

As the big dog barreled back along the fence, Puppy tried to turn around in midstride to run beside him and tumbled to the ground. I quickly stepped over to her, but she was back on her feet, grinning, before I got there. The neighbor dogs stopped running and came to the fence again to sniff, and this time Puppy trotted over to the fence and sniffed back, nose to nose. That made me chuckle and tell them all, "Good dogs! Good dogs!"

I heard my neighbor calling the bulldog and its big companion to their breakfast, and they were off like a shot. "Let's go inside too, Puppy," I said. She turned from the fence at the sound of my voice and walked along beside me up our sloping backyard to the house. The back porch steps were too high for her, so I picked her up and carried her inside.

In the kitchen I held the puppy in my arms and saw her eyes slowly closing and then blinking open. She had tremendous energy, but racing the older dogs had worn her out for the moment.

I put her down by the water bowl to see if she wanted a drink. She did, and I quietly repeated, "Drink, drink," while she lapped up some water. Thirst slaked, she was more than ready for a nap. I led her over to the new dog bed in the living room and encouraged her to lie down. I praised her as she did, and then contentedly watched her eyes close and her breathing deepen as she fell asleep.

She was a beautiful puppy inside and out. Her coat, more white than black, might not be the most favored for a Border collie. A mostly black coat stands out from the sheep better when the farmer is at a distance. But she was a pretty puppy — there was no doubt about that. The white parts of her coat had hints of mottled gray, which would likely become more pronounced as she matured. She had hazel to brown eyes, depending on how the light hit them. Her eyes shone, and I saw there her enthusiasm for life and interacting with Sally and me—and, I thought, keen intelligence.

Sitting beside her as she napped, I thought about the behaviors and characteristics she'd displayed so far. Her responsiveness to subtle changes in the tone of my voice, especially when I encouraged and praised her, her fast-increasing ability to maintain attention on me, her slight hesitation in chasing the third squirrel and her pulling slightly less on the leash during the rest of the walk—all these things suggested both that she had a quick mind for learning and that we were building a strong bond between us. What excited me even more was how quickly she seemed to be attaching meanings to "here," "out," and "do your business," associating the sound of each word or phrase with an appropriate action on her part.

The puppy was a long way from any referential understanding of these words. But after all my experiences with animals in and out of the lab, these signs indicated that she had enormous potential for learning. That gave me goose bumps.

Sally came out of the bedroom and I looked up to see her smiling at the puppy and me.

"You're up early," I said.

"I'm excited about our puppy," she said. "What have I missed while you've been up and about? You must have been doing something to tire this little dynamo out."

Sally knelt down and stroked the puppy. She stirred and stretched herself awake at Sally's touch, and then clambered into her lap and arms just as she'd done at Wayne West's farm the day before. After our years without a dog, it filled my heart with joy to see a puppy basking in Sally's nurturing glow again.

I told Sally about the morning so far. She approved of everything except my concern over the feral cat's apparent interest in the puppy. "Don't worry yourself about that," she said. "That cat's not going to get a chance to hurt our puppy."

But I wasn't convinced. "That cat's twice her size," I said, "and you yourself said a few days ago that it was wreaking havoc among the songbirds and squirrels."

"Hush, John. The birds and squirrels are tiny compared to the puppy." She held the puppy close to her face and said, "And you're gonna get big fast, aren't you?"

"I don't know," I said.

Sally kissed the puppy and put her down on the floor, then turned to me and said, "What about a nice breakfast on this special day?"

When I first met Sally she was a student nurse at Philadelphia General Hospital and I was a seminary student doing a counseling tutorial with one of the hospital chaplains. I was immediately captivated by her petite beauty, her oval face framed by short brown hair under the double-frill cap that Philadelphia General nurses wore, but also by the way she was joshing with a patient. This was in the psychiatric ward, and it moved me to see how Sally was giving an obviously very troubled man a few moments of relaxed fun. Her hair has turned white, but as throughout our marriage she can still josh me out of a troubled state of mind with her radiant brown eyes, her loving smile, and her unfailing common sense.

While Sally made scrambled eggs and I saw to toast, coffee, and orange juice, the puppy wandered around the first floor. The door to the basement stairs was closed as usual, and the stairs up to the loft area where guests stayed and I had my desk and files were a little too high for her. When she brought a toy over to one of us, she got a warm welcome and praise. Intermittently I verbalized actions she was making—"Take toy," etc.—to continue reinforcing the associations she was quickly forming with the sounds of those words.

When we sat down to breakfast, we gave the puppy her morning kibble. We wanted to build the habit of having her eat her meals when we did: breakfast, lunch, and dinner while she was a puppy, and then two meals a day, morning and evening, when she was fully grown. Sally and I grinned at seeing the puppy wolf down her food with gusto.

After we cleared the table I figured it was time for a little formal obedience training. If we couldn't reliably get the puppy to come to us when we called, it would never be safe to let her off the leash near a road and it would be hard to extend her learning in any dimension. So that would be her first lesson.

Teaching the puppy to come to the sound of "Here!" also had the virtue of being an easy lesson. I cut up some little pieces of cheese to use as lures and rewards. Even after a good breakfast, the smell of savory cheese got her attention. Then I knelt down in front of the puppy and held a piece of cheese a mere two inches from her nose while I softly said, "Here! Here!" She moved forward to take the cheese in her mouth and gobble it up, and I praised and petted her as she did.

Any complex behavior is a chain of simpler behaviors. In teaching the complex behavior it often works best to teach the chain of smaller actions in reverse and train from the end to the beginning. That way the learner, who could just as easily be a human student as a canine one, can progress through the chain of actions with increasing confidence, always secure in what the final desired result is. The idea of training from the end lay behind Wayne West's suggestion that a farmer with a young Border collie should let the dog out of the truck to follow its instinct and run around behind livestock after they'd been attracted by some hay. I used the same principle many times in experiments with rats, pigeons, and dogs in my lab at Wofford.

Step by step I began the process of getting the puppy to come to me and the cheese from farther away. From two inches we moved to four inches, eight inches, a foot, and then a few feet apart.

Five minutes of that was enough. I didn't want either of us to lose focus, and too much food at once would satiate her and lose its value for training.

Off and on through the rest of the day, we alternated training the puppy to come on hearing Sally or me say "Here!" with play with toys, exploratory sessions in the yard, and a couple of good naps. By the middle of the afternoon, the puppy came eagerly to "Here!" even when I was out of sight around a corner. Although she was delighted to get food treats, both yummy little biscuits and cheese, it excited me to see that praise and pets seemed to lift her spirits and please her even more than the food. That had something to do with both her individual temperament and her working Border collie lineage, I felt sure. In any case, it was another very promising sign for her long-term training. The food was an external motivation, and external motivations are never as strong and reliable as internal, instinctual ones.

Around four o'clock there was a knock at the door. It was our friend

Nora, a smile lighting up her face under her short, spiky blond hair. The puppy was enormously excited to meet another person. Ears up and tail wagging furiously, she lay down in front of Nora and looked at her expectantly. The puppy was so worked up that she wet herself and a small spot appeared and spread on the floor as Nora bent down to pet her and coo at her. A flower child with a country twang who was born and raised on a farm and had a dog and a cat of her own, Nora didn't bat an eye. We didn't make any fuss ourselves, lest we reinforce the urinating behavior. We knew this was a behavior that was likely to disappear as the puppy grew more accustomed to her surroundings and to meeting new people. Instead we just got a wet cloth to wipe the moisture off the puppy's fur and clean up the little puddle of pee on the floor.

Nora said, "I could have called, but I couldn't wait to meet your beautiful puppy. Do you think she's ready for a late-afternoon walk with the Ya-Yas?"

"I don't see why not," Sally said. "The usual time?"

"Yes, but don't you come by for us. We'll come around to you, so all the dogs can meet at once and we can get any hullabaloo over with."

Sally and a few of her women friends in the neighborhood loved the book and movie *The Divine Secrets of the Ya-Ya Sisterhood*. They dubbed themselves the Ya-Ya Winos and made that the tagline for their evening get-togethers, during which they imbibe a bit of wine and good Southern bourbon. The closest thing to a gang our neighborhood will likely ever see, Sally and the other original Ya-Ya Winos had first met years earlier while walking their respective pets. Based on their mutual love of dogs (to be fair, some members of the group have also had cats, like Nora's big fluffy orange cat, Slick), these crusty, savvy seniors — ninety-year-old Miss Lucy was the oldest, followed by Sally, Theresa, Nora, Marie, and Marge — soon developed a ritual of daily walks and twice-a-month get-togethers at one another's houses.

About an hour after Nora's visit, I looked out the window and saw her and two other members of the Ya-Yas, Marie and Theresa, approaching the house with their dogs. Nora had her Heinz 57 variety, Annie, a sweet swaybacked mutt with a penchant for eating dirt. Marie, a laid-back California transplant with light gray hair falling just below her chin, had her handsome golden retriever, Fafner. And Theresa had her miniature schnauzer, Holly. Tall with strawberry blond hair and porcelain skin, Theresa is a true Southern belle, and she was elegantly dressed for the walk as she is for every occasion. Sally put the puppy on the leash, and I went out with them to say hello to the ladies.

Marie said, "You should have gotten another dog ages ago."

I said, "Yes, we were just waiting for the right one."

"She's adorable," Theresa said.

"I told you all she was," Nora said.

Marie asked, "What are you going to call her?"

Sally said, "Nothing's struck us as fitting her just yet. Let us know if you have any suggestions. In the meantime, her name is Puppy."

Fafner, Holly, and Annie were milling around sniffing at the puppy. She sniffed back, but as with the dogs at the back fence she stood quite still and showed little interest in them. Nothing in her body language indicated fear or distress. Her ears and tail weren't down. But she was plainly much more interested in the people.

Introductions over, the walk began. The puppy didn't want to leave our yard. Sally beckoned to the puppy, gave a slight tug on the leash, and said, "C'mon, Puppy." But she didn't want to go. Sally wasn't about to put up with that, but she was too wise to engage in a contest of wills. I watched from our front porch as Sally scooped the puppy up. The puppy instantly relaxed in the cradle of her arms.

As Sally carried the puppy into the street I noticed the feral cat stalking toward them. Before I could say anything a car turned onto our block and the cat darted away. Sally turned back to the house and waved at me as she waited for the car to go by, and then the Ya-Yas trooped away on their constitutional.

At dinner Sally told me that she had carried the puppy most of the time. But on the last little stretch, the puppy had walked along on the leash very nicely.

"I'm still worried about that feral cat," I told her.

"That cat's doing anything to the puppy is unlikely, if you ask me."

"Well, I'm going to talk to the neighbors about not putting out food and water for it. Let it move on to some other neighborhood."

"John, you're really overreacting."

My voice rising sharply, I said, "Well, if you don't want me talking to the neighbors about the cat, maybe I'll have to call animal control about it."

Sally's voice rose to match mine as she snapped, "That's ridiculous!"

I was about to launch into a vehement reply when Sally said, "Where's Puppy gone to?"

The puppy had been lying quietly on the living room rug after eating her dinner. But now she was nowhere to be seen. Sally and I got up from the table and both called, "Here! Here, Puppy! Here, girl!"

She didn't come to us. A moment later we found her in the bedroom, lying at the far end of her crate. My heart sank at the distress our raised voices had caused the puppy, and at the thought that in one rash moment I might have undone the learning to come to "Here" that she'd achieved earlier.

Sally and I spoke softly to the puppy: "Here, Puppy; here, girl. Everything's all right. The storm has passed. Here, girl." A little hesitantly, she came out of the crate to be petted and praised.

"Gosh, Sally, I'm so sorry I spoke the way I did."

"I'm sorry about how I spoke too."

We both vowed to be more careful if we disagreed about something in the puppy's presence. Half an hour later, when we were all settled comfortably in the living room, I got a few treats and tried the "Here" exercise with the puppy again. Her first responses were a fraction slower than before the argument. But she was soon coming to find me eagerly, even if I was out of sight, when I said "Here!" in an encouraging tone. Fortunately no damage had been done, and I marveled at the puppy's extreme sensitivity to sound and tone of voice. That was a boon to training if I respected and capitalized on it, and a potential disaster if I ignored it.

A little later I got into bed mulling over the day's experiences. All in all it was a great day. We seemed to have a puppy as bright and quick to learn as she was loving. The Lobster King had bestowed bountiful gifts on her, and it was up to Sally and me to make the best or the worst of them. With luck, we would make the best of them.

But that feral cat still preyed on my mind.

5

"You've Got to Name Her Chaser!"

THE PUPPY'S OBEDIENCE training took on new urgency a month later during a visit from our young friend Allyson Gibson, a 2003 Wofford College graduate. Thank God Allyson came to see us that day. She ended up saving our puppy's life.

I met Allyson in her senior year during a January interim trip to the Everglades. Interims are a distinctive Wofford tradition. Between terms, students pick an intense month-long study that can have little or nothing to do with their major. Interim is a wonderful learning opportunity that is largely experience based, designed to get students out of their comfort zones and show them new ways of looking at the world.

Since I retired, Alliston Reid has invited me to help lead several interim trips. Alliston and I began traveling together on kayak trips and interims when he was an undergraduate, and it's been wonderful for me as a professor emeritus to join him and current Wofford students on interims in Florida and the Caribbean. It was on the second such interim in Florida that I met Allyson.

We were in the Everglades, off the grid, without the distraction of electronic devices. That provided ample time to get to know Allyson, who told me fascinating stories about spending her junior year abroad as that year's Wofford Presidential International Scholar. Studying the impact of globalization on world music traditions had taken her to India and other developing countries in Africa and South America. The experience had left her wondering if she should dedicate her life to being a medical missionary rather than pursue her passion for physics.

In the end she'd decided to enter the graduate physics program at Washington University in St. Louis. Sally and I were eager to hear about Allyson's first year in grad school.

Puppy had another idea. As we sat down in the living room, Puppy brought over a small ball and dropped it at Allyson's feet. Allyson delighted her by rolling the ball across the floor for her to chase down and capture. In between rolls, Allyson looked up and said, "Doc and Sally, I can't believe you've had this sweetie pie a whole month and are still just calling her Puppy. Don't you have any idea what you're going to name her?"

Sally laughed and said, "Everyone who meets her asks us about her name. But we want to give her a name that really expresses her personality, and we haven't figured that out yet."

I chimed in that names are powerful, and because we didn't want to name her lightly, her name was seriously eluding us. I had offered up "Baby," since I was calling her that anyway, but that was immediately shot down by Sally and the peanut gallery. Robin and Deb had not yet met her, but they were getting tired of hearing us constantly call her "the puppy" and "Puppy." They suggested a multitude of forgettable names from Girl to Lassie, Wilma, and Tasha (to go with Yasha). The list went on and on, and nothing seemed quite right.

"I am going to teach Puppy lots and lots of proper nouns," I said. "She has to have a name that fits that quest as well as her personality."

Over the past four weeks I'd spent more concentrated and extended time with Puppy than I ever had before with a young dog. Her personality and temperament were unmistakably soft, rather than rough-and-tumble, yet also — and without ever losing that softness — increasingly confident, curious, and even bold. Both aspects came out in her social behavior with people, as they just had in making a new friend and playmate in Allyson.

By the time Allyson rang our doorbell, Puppy had successfully interacted with dozens of people with growing assurance. She used an instinctive repertoire of behaviors — bowing her legs, looking up with her tongue lolling out of her mouth, licking the person's face, wagging her tail — that evolved in dogs in the wild and became a foundation for their unique social relationship with human beings. She already was adept at varying that repertoire to suit the individual. I had never observed a young puppy so persistent and creative in winning people's engagement, even if at first they wouldn't crack a smile. When we met a group of people she usually wanted to have some kind of interaction with each of them.

Dogs have evolved to be ready to interact with humans, and Border collies in particular have been bred to work in close harmony with farmers and their families. So Puppy's interest in people had a lot to do with her species and breed, but her rate of learning seemed remarkably fast. I had never devoted so much concentrated time to training a dog before, and I had to allow for that. But still, I found myself marveling at how smart dogs are. Our puppy seemed to have such a good mind, she was so eager to engage with people and so sensitive to voices, and her attention span was increasing so quickly. I was convinced she could learn that words have meanings and that names are labels that refer to specific objects, things, or individuals.

This is something that toddlers do. They first learn words, including their own names, by associating the particular sound of the word with a particular object or individual. A few months later they take a remarkable leap and come to understand that some words refer to categories. I was hoping Puppy could take that leap, and everything I saw in her made me believe that she could. Her fast emotional and cognitive development made me think we could soon ramp up the language training.

Puppy continued to show little or no interest in other dogs. The two dogs whose backyard abutted ours were friendly race competitors along the chainlink fence. And Puppy quickly accepted the dogs that belonged to Sally's fellow Ya-Yas. But she made no effort to play with them. She only wanted to engage with Sally or, if they showed any inclination for it, one of the other Ya-Yas.

Witnessing this behavior, I finally understood a comment of Wayne

West's: "A dog doesn't need another dog as a buddy. A dog needs a master or a mistress."

Sally and I preferred to think of ourselves as our dogs' parents and older buddies. We presented ourselves to Puppy as Pop-Pop and Nanny, the same names that our grandson, Aidan, knew us by. We happily indulged Puppy as we did Aidan, and as we had our daughters when they were young, but without spoiling any of them. To be a good daddy and momma to Puppy, we had to set appropriate boundaries for her and teach her to be happy within them.

Although the great Border collie breeders and trainers I know, Wayne and his friend David Johnson, generally do not speak of themselves as their dogs' parents, that's how I see them. They are tough-love daddies who foster their dogs' development with care and concern for them as individuals.

Working Border collies must have great confidence and skill in cooperating with people, including responding to verbal as well as visual cues. Their instincts give them the potential to do this. But they can only develop the confidence and skill they need through their relationship with their trainer.

Speaking very much as a parent to the dogs he trains, David told me, "Training dogs is kind of like raising children. I've got three grown sons, and I use the same approach with my dogs as I did with my boys. You just have to be consistent every day with them, children or dogs.

"Every dog that's brought to me for training, I have to find something in that dog that I like, and dwell on that and learn to love that dog. Because if I don't like the dog, the dog is not gonna like me. If I give my heart to the dog and can get the dog to like me and love me, then when he gives me his heart he'll give me his brain. When the dog does that, we can really accomplish some training."

Whenever Wayne had a litter of puppies on his farm, he and his wife made sure to interact with them positively and to provide opportunities for them to play with their grandchildren and other young children. David also testified to the value of interaction with young children, especially when puppies were in the critical early developmental phase from eight to twelve weeks of age. During these weeks, puppies are the most impressionable they will be in their lives, not counting their very first experiences with their mothers and litter-mates. It is a critical time for bonding with people as well as for learning other habits.

By the same token, Wayne and David shared with me how difficult it could be to train Border collies that had not had positive experiences with people. They spent enormous effort in building loving, trusting relationships with such dogs. The behavioral psychologists Clive Wynne and Monique Udell have shown that shelter dogs that have little or no contact with people are not readily able to focus on what humans are paying attention to and whether they are receptive to interacting with them. With some human attention, however, the dogs gain, or regain, this ability.

We didn't have any young children in the house, but as much as possible I played with Puppy like a little kid myself. I got down on the floor or the grass with her as we played with different toys, or to investigate something that attracted her interest. Within the limits of my human senses, I wanted to see what she saw, hear what she heard, and smell what she smelled.

Now Puppy had her focus squarely on Allyson. I watched Puppy bringing out Allyson's inner child, as Allyson alternated between rolling the ball straight at Puppy, letting her race in to get it, and rolling the ball quickly at an angle so she could race to cut it off. Each time, Puppy stopped on a dime when she got to the ball and picked it up in her mouth, looking proudly at Allyson while she held her trophy. Then she hesitated, as if she was debating whether she should give up the ball or not.

Part of this reluctance, I believe, was instinct — the longer Puppy had the ball in her mouth, the more it took on the character of food and the more reluctant she was to give up what felt like a yummy treat. But Puppy was also gaining confidence in Allyson's willingness to continue the game and wanted to exchange roles. "It's my turn now," her behavior said. When friendly dogs play, they frequently trade roles as they chase each other around. Occasionally Allyson obliged, chasing Puppy around our open living room, dining room, and kitchen area. Without any prompting on my part, Allyson always gave Puppy an energetic "Good girl!" for releasing the ball, and then rolled it across the floor again.

"Way to go, Allyson," I said at one point. "Reinforce her bigtime for giving you the ball, and then show her how that produces more fun chasing and capturing it again."

"Roger that, Doc," Allyson replied with a grin. She had participated in programs for prospective astronauts at the NASA Academy, and some of that lingo had become part of our banter.

Puppy exhibited classic Border collie behavior as she played with Allyson. The way Puppy attacked the ball instead of waiting for it to come to her typified the athleticism of the Border collie. Likewise, cutting off the ball at an angle was analogous to cutting off the escape of a sheep when it broke ranks and tried to peel off from the flock.

I was taking advantage of these and other Border collie behaviors to teach Puppy words by association. When I threw a ball into the middle of the floor or the backyard, she instinctively circled behind it, like an adult Border collie circling behind some sheep or cows. I said, "Puppy, come by," if she went clockwise around the ball, and "Puppy, way to me," if she went counterclockwise. In the course of chasing and herding play with the ball, I voiced all the commands that farmers give Border collies — go out, come by, way to me, there, drop, crawl, look back (there are livestock behind you), that'll do (stop what you're doing and rush to the trainer or shepherd) — by labeling her spontaneous behavior out loud. I also wove the basic obedience signals — sit, stay, here — into our play. The hope was that tying the obedience behaviors and their names into her instinctive behaviors would speed up the obedience learning and make it easier to direct her in chasing and herding and other activities, including word learning.

The process was working out beautifully. All things considered, Puppy was developing fast. And I wasn't shy about bragging on her.

"Puppy is learning to come, sit, stay, and all that super quick," I told Allyson.

She asked for a demonstration, so I put Puppy on the leash and we all went outside. Sally sat on the porch to watch the show while Allyson and I took Puppy into the front yard.

It was almost noon and the late-July sun was beating down, but the big trees around our house provided large patches of shade. Before I dropped Puppy's leash, I glanced around to make sure the feral cat was nowhere to be seen. I'd glimpsed the cat often recently, slinking down into the gulley toward the drainpipe, and I'd about decided that Sally was right and I was overreacting.

Until the day before, when Puppy had come to a halt right near the end of our early-morning walk, a few paces from our driveway. Without looking to see what had stopped her, I gave a gentle tug on the leash. She didn't budge. I turned and saw her frozen in midstride, body taut with excitement, her head and shoulders dropping into an instinctive stalking posture, and her eyes locked on something to the right.

I followed her gaze, and dammit if it wasn't that pesky coon cat, not four feet away and staring right back at Puppy. The cat was perched on a sawn-off tree trunk, arching his back so high that I could see the just-rising sun under his belly. His tail pointed out behind him, snapping back and forth, and his entire body was puffed up to almost double his real size. His eyes burned into Puppy's as he opened his mouth wide and expelled a long, ugly hiss, followed by the loudest, meanest growl I've ever heard come out of a cat.

That was all I needed. I scooped Puppy up in my arms and boogied on down our driveway and into the house. As soon as I entered the door I hollered, "Saaaaallllyyy!" Once again, however, Sally did not put much stock in my fears.

Thankfully today there was no sign of the cat. Allyson and I took Puppy into a shady spot and I dropped the leash.

I worked up close with the puppy at first, and then gradually increased the distance between us, keeping her in the shade as much as possible. I asked her to sit, lie down, stand, stay, and come, and she did so happily. Tongue lolling out and tail wagging, she delighted in the praise, pets, and treats I gave her for her good performance. And she was exultant in the moments of play with Allyson and me that we interspersed here and there.

I gave Allyson a few little dog biscuits, and Puppy went through the obedience behaviors almost as readily for her as she did for me. Then I asked Puppy to sit under a big pine tree at one side of the yard. While Allyson stood watching from the front of the yard, about midway between us, I retreated to the opposite side, a good sixty feet away. It was quite a distance for a twelve-week-old puppy to keep her focus on me and my words. The thought crossed my mind that I was being overconfident, but I dismissed it. Everything was going great.

I was just about to call Puppy to me when a red Jeep Cherokee with the windows blacked out barreled down Seal Street beside us. Traffic was usually particularly slow on Seal Street, because it is only about seventy yards long. We had purposely not yet taken the puppy near any busy roads, so she had not yet seen a vehicle moving fast except when we were driving in our own car.

The instant Puppy heard and saw the Jeep she was on her feet. As it whooshed by she skedaddled after it despite my cries of "No! No!"

From the porch Sally yelled, "John! Get the puppy!" Allyson was the closest and she had already sprinted in pursuit, kicking off her flip-flops and dashing onto the asphalt in her bare feet. Sally rushed down the porch steps after her. The farthest away, I was also the slowest to react. I strained to catch up as the three of us chased the puppy chasing the red Jeep Cherokee toward Briarwood Road.

We all frantically called out, "Here, girl! Here, Puppy! Come!" But she kept going as fast as her little legs could carry her. She was lost in the excitement of pursuing the big red sheep. What was scaring the bejesus out of us was that, in classic Border collie style, Puppy saw that the Jeep was turning right onto Briarwood and she was cutting across the neighbor's lawn to head it off. She was in danger of running directly into the front wheels of the Jeep as it accelerated out of its corner-hugging turn.

I saw the blur of the red Jeep and Puppy racing toward it. Completely panicked, sprinting with all my might, I reached the corner as Allyson dove forward and grabbed Puppy's leash. Allyson hit the grass in full stretch and, just in time, held Puppy back, sprawling her on the grass too as the Jeep sped away.

Sally was in tears as she scooped up Puppy in her arms. Finding her voice before I could find mine, she sputtered out, "Oh my god, thank you, Allyson. Thank you, thank you." I repeated the same, still shocked as the puppy squirmed out of Sally's arms back onto the grass.

The grin on Puppy's face, her pricked-up ears, and her fiercely wagging tail showed that she was completely pumped up about chasing the Jeep, and knew in the depths of her being that she had done exactly what she was bred and born to do. She was bursting with exhilaration and pride in her exploit. It was too late to administer a correction for the unwanted behavior. Any fuss on our part would only make the incident more memorable for her.

We helped Allyson off the ground and hugged her tightly. I gently took the leash from her, and quietly encouraged Puppy to walk back to the house with us, trying to regain my composure as Sally laced her arm through Allyson's. My heartbeat gradually slowed down after ten of the worst seconds in Sally's and my life.

A few minutes later we were all having lunch together, laughing with Allyson about her flying grab of the leash. Sally and I told her we were lucky her instincts were as strong as Puppy's. After finally telling us about her good first year in grad school, Allyson bid Sally, me, and not least of all Puppy an affectionate farewell.

Later that afternoon we had more excitement with an impromptu visit from Robin. Impatient to meet the puppy she had been hearing about for the past four weeks, and on a break between guiding whitewater rafting trips, she had decided in characteristic fashion to drive all the way down from her home in the mountains of western North Carolina and surprise us. We gathered in the kitchen, giving Puppy lots of pets and soft words as we filled Robin in on events since our last conversation on the phone.

Tanned and toned from her long hours on the river (she can beat most men in a push-up contest), Robin was standing at the sink drinking a cup of coffee. When she heard about the puppy and the Jeep Cherokee, she doubled over with laughter and almost dropped her cup.

"You've got to name her Chaser!" Robin said. Sally and I exchanged looks of instant approval and laughed along with Robin, marveling at how we could have missed the obvious. We said we wanted a name that reflected our puppy's inner self, and all the time it was right there in her behavior. As Joseph Campbell, a big influence on me, might well have said, our puppy was a veritable avatar of chasing. I instantly imagined her chasing speech as well as sheep.

Chaser felt like exactly the right name to me. But I wanted to be sure we all agreed. So as I always liked to do in class, I went around the room and asked for opinions.

"Sally, what do you think of the name Chaser?" I began.

"I really like it."

"Robin, are you having any second thoughts?"

"No, Dad, I still like it. What do you think?"

"Well, I really like it too. But we gotta call Debbie on this and see how she feels about it. Everybody's gotta like it."

"What do you think?" I said to the puppy as she brought a ball back to me. "How do you like the name Chaser? Should we call you Chaser?"

The word "chaser" meant nothing to her, of course. It was just an utterance of ours with two linked sounds. But at least she didn't find the sounds unpleasant, judging by her continued happy focus on play with the ball as we each tried out her prospective name. It seemed like a name she could learn to like.

Before Robin left we called Debbie in Brooklyn, and the vote was unanimous. Our puppy was now Chaser.

That evening Sally and I reminisced about how Robin and Debbie had named our dogs going back to Bimbo, the big German shepherd mix we got when they were in the middle of elementary school. Bimbo's name fit his rambunctious yet goofy personality, which made him a great companion for Robin and Debbie in their adventures. After Bimbo came Yasha and Grindle, both perfect names in their own way. Yasha, half Border collie and half German shepherd, had the heart of a Cossack warrior chief and the temperament of an imperious virtuoso. Grindle was named ironically for Grendel in *Beowulf*, but creatively misspelled by me. In no way a monster, Grindle was a beautiful big purebred German shepherd with the softest of hearts who was fiercely devoted to the whole family, especially Sally.

It was good to have our new family member's name. But chasing the

Jeep had excited her so much that I had to devise a lesson memorable enough to counteract it. We couldn't risk another incident like the one we had just experienced.

The speeding Jeep was a blessing in its way. It demonstrated without a doubt that Chaser had a strong herding drive. Herding is more than chasing, of course, but the underlying behavior for herding is chasing. Add chasing in a designated direction, and you begin to get the countless variations and possibilities that make up a working Border collie's responsibilities and problem-solving challenges.

On the other hand, chasing cars was an all-too-lethal challenge. In my mind's eye I kept playing back the nightmare moment when the Jeep turned and Chaser raced to cut it off. Fortunately, she failed in her attempt. But it was too close a call.

As I watched Sally playing with Chaser, I recalled Wayne West's telling me about refusing to sell some people a dog because they didn't have a fenced-in yard. Wayne said, "If they ain't got the facilities to keep a Border collie, I don't sell them a dog. I tell them, 'This dog won't last two weeks. It'll get run over. Whether it's a car or a bumblebee, this dog's gonna try to work it and herd it. And the car is the most challenging and exciting thing for the dog to pursue. The dog's gonna be out there running after the car. And if a dog gets started doing that, it's hard to break them of it.'"

Tomorrow I would have to redirect Puppy's powerful impulse to chase, without quashing the impulse itself. I couldn't let her continue to have an unbridled instinctual desire to chase cars. But sure as shooting, I did not want her to lose the instinctual joy of chasing, either.

6

Chaser Learns What Not to Chase

NOW I HAD two problems to deal with: chasing cars and that darned cat.

Driving back from Wofford the next day, I had time to stew about the threat of the cat. I was completely flummoxed by Sally's casual attitude, particularly after the last run-in with the cat. Both my eyes and my gut told me that the cat was stalking Chaser. She was less than half his size, so Chaser was no bigger challenge for him than a squirrel. As soon as I got home, I would insist that we call the animal shelter. Working myself up, I rehearsed my argument out loud. Maybe I was being a little childish, but the cat had to go.

When I hurried through the door, Chaser ran to meet me, wiggling and squirming on the floor for my attention like a fuzzy little inchworm. I couldn't help but smile, and the tension in my back and shoulders started to fall away as I knelt down to scoop her up in a hug.

Sally came into the living room and walked up to me and gave me a kiss before I could get a word out of my open mouth. I took a breath and began again, feeling much less agitated than I was in the car. "Sally, I need to talk to you about . . ."

"Okay, hon, just a minute," Sally said. "Before I forget, I want to tell you I've taken care of the cat."

Stunned by that, I asked, "What do you mean, sugar?"

Sally explained that she had been picking the brain of Lynn, our neighbor three blocks over and an avid cat lover, about the best way to deal with the feral cat. Lynn's first suggestion was that we adopt the cat. Sally confessed that she thought that would be unfair to the cat with all of our attention showered on our new puppy. Lynn said she completely understood and would put some thought to it.

While I was out that morning, Lynn had come by the house with a cat carrier. She and Sally had walked over to the drainpipe opening in the little gully across the street, and Sally had called, "Kiiiittty! Kitty, kitty, kitty!"

The cat had come shyly out of the drainpipe, meowed, and rubbed up against Lynn's legs. With a little cheese, they had lured it into the carrier, and the feral cat was now a domestic cat, living with Lynn and her husband, Ken.

I wrapped Sally in my arms and said, "Thank you, thank you, thank you, sugar. You are amazing!" Chaser had to get in on the hug too, wiggling and squirming against our knees with playful puppy yelps.

This eliminated one of my two fears. But that still left chasing cars — or anything else that brought Chaser running out into the street.

I was blown away by Chaser's speed in pursuing the Jeep. I knew instinct was a force to reckon with, as I'd observed many times in animals over the years, in and out of the lab. But Chaser had just given as dramatic an example of instinct's explosively powering a behavior as any I'd ever seen. Besides that, I knew that the thrill of racing after the Jeep would almost certainly enhance her desire to chase tons of steel on wheels. It gave me a shudder to think that my overconfidence had put her in danger.

Instinct is powerful stuff. The release of an instinctual behavior is inherently self-reinforcing. And the more memorable and exciting a behavior is, the more likely an individual will be to repeat it. That was what made the Jeep incident so troubling. It was the most exciting experience in Chaser's young life and thus very positively reinforcing. She was going to be vulnerable to repeating the behavior of chasing cars unless I took effective action.

At the same time, I didn't want to quash her instinct for chasing. I was already making use of that instinct to teach Chaser the names of the objects we played with every day. I wanted to get the full energy of her chasing instinct—the energy that had startled and even frightened me with its intensity the day before—focused on every detail of that play. So I couldn't be heavy-handed. I had to channel that instinct carefully, as gently as possible.

The modern concept of instinct goes back to Charles Darwin's theory of evolution. However, the study of instinct lay dormant until the 1930s, when European zoologists began to study the behavior of animals in their natural environments. This was the birth of a new science of animal behavior called ethology. The founding generation of ethologists and their immediate successors cast new light on instinct. They established that instincts have survival value or else the instincts themselves do not survive; that all members of a species have the same instincts in common, although the strength of a particular instinct varies from individual to individual; and that the instinct maintains its essential features over the animal's life span. Compared to a reflex, such as your lower leg kicking after the doctor taps your knee, an instinct is infinitely more complex and may consist of a series of behaviors.

Ethologists have shown that an instinct consists of two major components, a fixed action pattern and a releasing stimulus. The energy for the fixed action pattern is stored up in the animal, while the releasing stimulus, as the name implies, merely releases the energy. The fixed action pattern and the releasing stimulus are like a stick of dynamite and a fuse. The fire of the burning fuse is nothing compared to the explosive power of the dynamite.

The Dutch ethologist Nikolaas Tinbergen began studying that explosive power as a child by observing the mating and nesting behavior of two-to-three-inch-long stickleback fish in his backyard pond. Fortunately he continued that study as an adult, and in 1973 he was awarded the Nobel Prize in Physiology or Medicine for research that included the instinctive mating and reproductive behaviors displayed by male stickleback fish.

There comes a time when the male stickleback fish selects an area as his territory. If another male appears, the first male attacks furiously and always wins. Tinbergen concluded that energy for fighting is built up in the first male, just waiting to be released, and this built-up energy enables him to defeat the intruder. Having defended his territory, the male stickleback builds a round nest with sticks and weeds. When a female stickleback fish with a swollen belly appears, the male performs an elaborate mating dance to lure her to his honeymoon nest. Once the female enters the nest, the male noses her tail, and she ejects thousands of eggs for him to fertilize.

Like all instinctive behaviors, the male stickleback fish's mating behaviors fly under the radar of consciousness. The male stickleback fish also illustrates that instinct is often linked to critical developmental and maturational stages in the life of an organism. If something hinders normal expression of the instinct during these critical stages, it may not achieve full, or even partial, development.

We know that there is a critical period for social development in young dogs. I suspect now, as I did then, that there may be a similar critical developmental period for developing and channeling the herding instinct. Sally and I were trying to channel our puppy's chasing and herding instincts into pursuing balls, Frisbees, and other toys. For Chaser, as for all dogs, the stimulus for her chasing instinct could be anything moving, such as sheep, rolling balls, flying Frisbees, squirrels, cats, and Jeeps. Much of what I'd been working on with Chaser during the past month was providing safe stimuli for chasing in the confines of the yard, with the goal of neutralizing the effect of dangerous stimuli like squirrels, cars, and big feral cats that could draw her into the street. But honestly, balls and Frisbees were poor substitutes compared to cars and critters.

A simple solution was available: confine Chaser to our fenced-in backyard with walks only on her leash. I quickly rejected that option. I wanted to enlarge Chaser's world, not shrink it.

I recalled reading that having someone in a moving car squirt water on a dog with a water-blasting toy, or throwing tin cans to clatter in front of the dog, can inhibit car chasing. However, the mental image of Chaser's cutting across a corner to head off the Jeep led me to reject both of these methods as too dangerous.

Somehow, some way, Chaser had to learn that chasing a ball or a Frisbee in the yard was different from chasing a squirrel, a car, a ball, a Frisbee, or anything else in the street. There were two means I could use to influence Chaser's behavior, positive reinforcement and negative reinforcement. All of my training thus far had emphasized positive reinforcement. But something inside me said that we needed to employ a few negative reinforcement procedures now in order to protect her fully and permanently from literally leaping into danger.

I've already said a little about positive and negative reinforcement. Negative reinforcement is certainly not punishment. The essence of punishment is that there is no way to escape or avoid it. By contrast, negative reinforcement motivates a learner to act in a way that escapes or avoids anything uncomfortable or dissatisfying, such as learning to carry an umbrella to avoid getting wet in the rain.

Sometimes positive reinforcement and negative reinforcement are both necessary to change a behavior. In the early 1970s negative reinforcement began my effort to stop smoking. Positive reinforcement completed the process, enabling me to stop smoking and never miss it.

I started smoking in my senior year at Abilene Christian College. The habit strengthened in graduate school at Princeton Theological Seminary to the point where I couldn't study without smoking. I couldn't concentrate without a cigarette in my hand.

Robin and Debbie hated my smoking and often complained that I smelled of tobacco even when I didn't have a cigarette going. When they were in elementary school they began to roll down the car windows in the middle of winter, no matter how cold it was, dramatically coughing and gagging at the unpleasant smell. To avoid the negative reinforcement of the blasts of cold air in the car and the girls' frequent distress and disapproval, I decided to stop smoking.

A month later I was drinking beer with some students in a local restaurant when one of them offered me a cigarette. I said I'd quit. The student asked me how long it had been, and I told him I'd gone a month without smoking. He laughed and said that he'd quit smoking several times, once for as long as six months, but he always wound up going back to it. I asked him if he ever got to the point where he didn't think about smoking. When he said no, I said, "Give me a cigarette."

Robin and Debbie were persistent, and I decided to try to quit smoking again a few months later. I confess another form of negative reinforcement also motivated me: smoking was decreasing my physical endurance and cutting down on my kayaking and windsurfing fun. Still, I knew I also needed positive reinforcement to motivate me to stay with it and succeed. I had been reading about a new style of rubberized, inflatable canoe that was coming on the market. I couldn't justify spending money on something so selfish, rather than something for the family, but I told myself I could buy the canoe if I used money saved by not buying cigarettes.

Like a kid saving for a new bike, every morning I put the money for a day's cigarettes — I was smoking about a pack a day — into a jar on the kitchen table. Whenever I felt the urge to smoke during the day, I added a little more money to the kitty. In the evening I dumped all the coins out on the living room floor to count them. This always engaged Robin and Debbie, who eagerly helped me count the money and then cheered and did a little dance to encourage me to stay steady on my path. This became a daily after-dinner ritual for the whole family, and we were all looking forward to the new canoe together. My dream became our dream, which made it all the more motivating.

I also conditioned myself not to think about smoking. My initial goal was not to think about smoking for thirty minutes. If I managed to do that, I rewarded myself with five minutes of visualizing running rapids in my kayak. Slowly, I extended the time not thinking about smoking to forty-five minutes, an hour, and longer periods.

After a couple of months I got so good at not thinking about smoking that people could smoke in front of me and it didn't affect me at all. Smoking was traditionally allowed in the college. I had smoked frequently in class, and students did too. Coincidentally, the college at that time decided to ban smoking in classrooms, and the chairman of the Psychology Department told me, "Pilley, you've got to stop students from smoking in your class."

"They're not smoking," I said. When he insisted that they were, I realized I'd gotten to the point where I didn't even notice when people were smoking around me. To top it off, I could walk up multiple flights of stairs at Wofford without getting out of breath. Around the same time I had saved up so much cigarette money that I could buy the inflatable canoe. Lovingly dubbed the Rubber Ducky, it was virtually impossible to capsize and became a favorite of the entire family, especially the girls.

On the basis of my successful effort to quit smoking, I concluded that as I had to quit smoking, Chaser needed a little negative reinforcement, as well as lots of positive reinforcement, to quit chasing cars. Negative reinforcement, often in an implicit form, is usually part of healthy boundary setting. The question was exactly how to provide this.

The experience of negative reinforcement depends on an individual's temperament and personality. Some individuals are from birth thicker-skinned or more hardheaded than others. Trainers often talk about hard dogs and soft dogs. Yasha was a hard dog, and in his early obedience training he needed a fairly strong, but still gentle, hand. Grindle was a soft dog. Observation over the last month of Chaser's equally soft temperament suggested that a little negative reinforcement should go a long way with her.

Before she chased the Jeep, I had already started to use some very mild negative reinforcement by adding a firm "No!" just before she ran to the end of her leash while chasing something. The idea was to associate the "No!" with the unpleasantness of hitting the end of the leash, so that the word would eventually stop her from pulling and lunging on the leash, and also generalize to become a stop signal for other unwanted behavior. But unfortunately the jerks she was giving herself were not sufficiently aversive to give the "No!" real strength. Trainers often suggest that when dog creates a jerk on the leash, we should intensify it enough so that the dog *really* doesn't like it. Otherwise the dog will make a judgment such as, "If I lunge after the cat, I will get a tug. But it's really not that bad, and it's worth it because I love to see that cat run."

I finally made the tough love decision that to enhance Chaser's immediate responses to my commands, I would take her into the front yard on a fifteen-foot line, two and a half times as long as her normal leash. The long line would let her build up more speed in chasing something she shouldn't, and she would then get more of a jerk when she ran to the end of the line. I also planned on adding a firm tug of my own to the long line to heighten the jerk.

When we went outside, however, I couldn't bring myself to take that step. It seemed too abrupt. Instead, I took her into the backyard to rehearse the major obedience commands with no possible distractions from passing cars — and using only positive reinforcement. I made sure Chaser knew that I had treats in my pocket. Without positioning her, I wandered around the yard. Every minute or two, I randomly gave one of the obedience commands — there, stand, drop, crawl, here — in combination with "Chaser!"

Every time she heard her new name, Chaser instantly looked at me and I paused only a second before giving an obedience command. I smiled as she eagerly responded to my words. I lavished yummy treats, praise, and pets upon her with enthusiasm, going through several repetitions for each of the obedience commands.

A few hours later, I again took Chaser into the backyard and ran through the same exercises, again using only positive reinforcement. At the end of the session, I switched over to the obedience herding game, using a tennis ball as a surrogate for a sheep. In the course of the game I gave Chaser the obedience commands relative to the tennis ball, hoping that the herding scenario would cause the obedience commands to take on even more positive value. The success of these training and play sessions made me think that maybe we could still avoid using negative reinforcement beyond "No!"

That evening Sally and I were sitting on the front porch with Chaser lying quietly between us when a car went by, not as fast as the Jeep but still moving quickly. Chaser instantly got to her feet and barked at the car. I grabbed her collar to prevent her from scrambling off the porch in pursuit, and we took her inside.

We had to do something more. But it had to be something that wouldn't throw her off stride in other ways. One firm jerk on the long line would surely be okay, given our good relationship with Chaser and her abundant confidence. But having to repeat that a lot worried me. Stroking Chaser, Sally said, "If the momma dog was here she'd probably get you to stop doing something she didn't like with one little shake on the scruff of your neck."

"That's brilliant, Sally! That's what we'll do," I said.

Puppies are biologically prepared to learn that a behavior is not allowed when the mother dog shakes them by the scruff of the neck. We knew from Wayne West that he relied on the mother dog's doing that to teach young puppies not to defecate and urinate in the kennel. Chaser had just turned twelve weeks old and was still at her most impressionable. If I did things right, one lesson about going into the street might be enough.

The next morning I took Chaser into the front yard on the long line. Luck was with me and it wasn't long before a car came by. Chaser immediately headed after it. As she was building up speed I said, "Chaser! Here!" She kept going. Just before she got to the end of the fifteen-foot line, I roared, "No!" When she hit the end of the line, I gave a firm tug back on it. And then I ran to her, picked her up by the scruff of the neck, and gave her one vigorous, mother-dog-like shake.

I gently put her down on the grass, moved a few feet away, and in a warm, soothing tone said, "Chaser! Here!" She came immediately to be comforted and given a treat.

Over the rest of the day's normal routine of play, obedience training, and walks, Sally and I found that the "No" was enough to stop Chaser in her tracks if she ran after a squirrel or headed into the road for any reason. By the end of the day the "No" did not even have to be shouted, although we always made sure it was firm and clear and loud enough to be heard.

There was just one more test we needed to complete Chaser's don't-go-into-the-road training. The following morning we all trooped into the front yard — with no long line or leash. I told Sally I wanted her to walk into the road, luring Chaser to follow her. Sally doubted that Chaser could be tempted from my side, but she obliged me and walked toward the road. Curious, Chaser followed her. As Chaser neared the edge of our yard I said, "Chaser, here!" She kept going, and worry rose up in my head. But just before she got to the road, I said, "No!" Chaser stopped instantly and looked back to me. I called out, "Chaser, here," and she came to me without hesitation.

I gave her all the positive reinforcement I could muster, and Sally quickly joined us and did the same. A little later we switched roles, and I went into the road while Sally gave Chaser her commands. Again we found that "No" now stopped Chaser from going into the road.

That was the turning point. Over the following week, we strength-ened the power of "No" through repetition. The word had now ac-quired aversive meaning for Chaser. And once "No" took on a quality Chaser didn't like, it became something she really didn't want to hear. With consistent use, the "No" would lose its negative emotional qual-ity, but still remain effective as a command cue and a directive piece of information. We could even use "Yes" and "No" as substitutes for "Hot" and "Cold," when I asked Chaser to find a hidden object.

An additional tactic we used to teach Chaser not to chase the wrong things was to positively reward a response that was incompatible with chasing. When we saw or heard a car coming during a walk, we pointed to the grass and said, "Car. Go to grass." And then we ran with her to the grass, had her lie down, and showered her with praise and pets un-til the car passed. At that point we softly said, "No car" or "Okay," and continued on our walk.

Within a few days, Chaser chose to leave the road and lie down in the grass whenever she heard a car coming, usually well before we did. We gave her lavish positive reinforcement the first times she did this. Soon our positive reinforcement could be more moderate without the behavior weakening. In fact, if we forgot to say, "Okay, Chaser, c'mon," after a car passed, she remained lying in the grass until we did. Sally was the first to experience this. Lost in conversation with the Ya-Yas, Sally suddenly realized Chaser was not with her. Spinning around, Sally saw that our bright little puppy was still patiently lying in the grass at the side of the road, waiting for the "Okay" command.

Thanks to the combination of these techniques, Chaser would not go into the street even to follow her favorite ball, unless I went into the street and asked her to join me there. This learning held in relation to any road anywhere we took her after that, from Spartanburg and envi-rons to visiting Debbie, Jay, and Aidan in Brooklyn.

Now I could concentrate on channeling Chaser's powerful social and herding instincts into her language learning. When it comes to harnessing the power of Border collies' instinctive drives, we can learn a lot by listening to the farmers and trainers who work with them every day.

7

—

Listening to the Farmer

THERE WAS A lot to be thankful for as Chaser turned four months old. The danger of the feral cat had passed, she had learned to be safe from cars, and most of all she had bonded strongly with Sally and me.

We positively reinforced Chaser's bond with us in every way we could. But reinforcement is definitely a two-way street, and Chaser was positively reinforcing us as much or more. Just as a smile elicits a smile, Chaser's playfulness stirred the same in Sally and me. Her wagging tail sparked joy in us to match her own.

Sally and I had been looking forward to having a dog in the family again. But we weren't prepared for how strongly Chaser affected us. We didn't realize how big the void in our lives had become.

We went puppy mad. We e-mailed dozens of pictures of Chaser to Debbie — Robin's e-mail connection in the mountains was too slow for that — and our conversations with both Robin and Debbie were full of details about our new family member. Debbie joshed with us on the phone one day, "I get it. She's cute." But like Robin, Debbie listened patiently as we waxed on and on about Chaser.

Perhaps the impact was even greater on me than on Sally. My mind was buzzing with curiosity about Chaser's potential to learn the meaning of words. Eight years after my retirement from Wofford I was fully a scientist again, entranced by the possibility of discovering something worth sharing with the world. Sally told me more than once during our first month with Chaser, "You've needed this, John."

It went even deeper than my desire to contribute to scientific understanding. The day we brought Chaser home, science became secondary to her quality of life. Not that there was any conflict between the two. The more Chaser fulfilled herself as a dog, a Border collie, and a member of our family, the more she was likely to be able to learn. But if a conflict between Chaser's needs and science ever came about, Chaser had priority.

Later in the conversation in which she joked that she'd seen more than enough pictures of Chaser, Debbie said, "I think she's given you your heart back, Dad."

I couldn't say anything. A few seconds later, obviously concerned about how I might be reacting at the other end of the line, Debbie broke the silence and said, "It just feels like ever since Yasha you've been holding something in. And now that part of you is breathing again."

I still couldn't say much. But I managed to voice my agreement and my thanks for her, Robin's, and Sally's love and understanding. The black box into which I had locked my grief over Yasha had opened without my realizing it. My sorrow over his loss and my lingering sense of guilt for his suffering at the very end were gone. Yet his vital spirit was still with me, stronger than ever, now that Chaser was a member of our family. My friend who told me, "When you get a pet, sooner or later you get a broken heart," also told me, "Your heart gets whole when you can risk its being broken again."

I hadn't been able to take the first step onto that path of emotional healing, but Sally had taken it for me. And then Chaser, greeting me every morning with a "What are we going to do today, Pop-Pop?" tail wag, had done the rest. Her arrival in our family was a blessing, for sure.

My scientific hopes for teaching Chaser language grew larger the more she displayed strong Border collie traits and instincts. Dogs of all breeds and mixtures of breeds can be highly intelligent, but no dogs have shown greater skill and creativity at problem solving than Border collies. The stewards of Border collie intelligence over the generations — farmers, breeders, and trainers — have amassed a body of understanding and insight that complements animal science and offers many clues for researchers to follow and test.

Scientists are trained to be skeptical about so called anecdotal evidence, meaning observations from experience that haven't been confirmed in rigorous experiments. However, the knowledge of Border collies' intelligence that those who live and work with them every day have accumulated goes far beyond anecdotal evidence. One of the greatest animal scientists of recent times, John Staddon, the James B. Duke Professor of Psychology and Professor of Biology and Neuroscience Emeritus at Duke University, believes that the knowledge of farmers and shepherds provides compelling evidence of dogs' ability to make inferences and solve complex problems without human direction. In Staddon's view the exploits of Border collies epitomize creative learning as opposed to rote learning.

In a 2006 essay for a special issue of the *International Journal of Psychology,* Staddon wrote, "There are two methods to train a dog" — and by extension a person or any other animal. The method that is "the quickest, least dependent on individual aptitude . . . is called 'shaping by successive approximations.' It is the method used by circus trainers. . . . It is effective and reliable, especially if what is to be taught is well-defined and predictable."

For example, a dog act in a circus might involve dogs jumping through hoops. Shaping by successive approximations begins with getting the dogs to walk through hoops that are stood up on edge on the ground or raised up only a few inches. Gradually the trainer brings the dogs to the point where they leap from one tall platform to another through hoops of blazing fire, perfectly on cue with the gestures of the trainer and any music or sound effects.

These are amazing feats of training, athletic skill, bravery, and, it should be said, intelligence. A circus dog must be a good learner and periodically be able to learn new tricks. It takes a strong bond of trust between trainer and dogs, and dedicated effort by all of them, to put on a great show. But the outcome, if all goes well, is predetermined. The dogs will never be asked to solve a problem on the spur of the moment by applying their instincts, life experiences, and reasoning powers to draw inferences. The trainer and the dogs know exactly what to expect during every microsecond of the act. There is no uncertainty or surprise, except for those in the audience who haven't seen the performance before.

Shaping by successive approximations is essentially closed-ended. Applied to humans, Staddon says, this method "is the basis for regarding teaching [children] as training in a 'skill,' like a trick to be taught to an animal. It treats students like dogs, and pretty dim ones at that — Odie rather than Lassie." This is what is happening today in schools where teachers are regimented into "teaching to the test," and where students spend more classroom time taking practice standardized tests than they do interacting with their teachers and classmates on curriculum content.

By contrast, "creative teaching" is required for teaching things that are less well defined and do not have a predetermined outcome. In the case of humans and animals alike, Staddon argues, this involves creating an environment that challenges learners to use all their "natural propensities"— instincts, fundamental drives, emotional energy, perceptual and cognitive abilities — to solve new problems.

It's important to realize that nature is the greatest creative teacher of all. It's equally important for us twenty-first-century humans to realize that we are not nature's only brilliant students and that we are not its only thinking species. Over the past decades, naturalists in the field have discovered attributes, abilities, and behaviors once thought to be "uniquely human"— tool use, "cultural" differences between groups that belong to the same species, complex emotions such as empathy, inferential reasoning, among other characteristics — in animals such as bonobos, chimpanzees, crows, dolphins, and not least of all domestic dogs.

Researchers are confirming and probing these characteristics both in laboratories and in animal sanctuaries that seek to replicate wild animals' natural habitats. For example, Brian Hare and Juliane Kaminski have separately devised and conducted experiments with bonobos, chimpanzees, and dogs strongly indicating that all of these animals have an implicit theory of mind comparable to that of young children when they acquire language. Hare's book *The Genius of Dogs,* written with his wife and co-investigator Vanessa Woods, offers a fascinating guide to their own work and that of many other researchers on the perceptual and cognitive traits humans share with animals. Much of this work draws on techniques first developed in behavioral and cognitive psychology experiments, with behavioral psychologists such as John Staddon pioneering a broader understanding of the role of inference in nearly all animal learning. The ability to acquire language, I am confident, will sooner or later be recognized as one more capacity that is no longer "uniquely human."

Staddon reaches the same conclusion that I had heard from Border collie breeders and trainers: creative teaching builds on the learner's instincts and innate tendencies. Instead of teaching single, repeatable behaviors in isolation or in a series, this method of teaching and training encourages creative learning, the ability to solve unexpected problems through spontaneous trial-and-error inference guided by accumulating experience and judgment. Creative learning is essentially open ended. It is also highly dependent on the relationship between the learner and the teacher or trainer.

Staddon offers two examples. One is of a teacher encouraging a child's interest in scientific discovery. The other is of a shepherd teaching a Border collie to herd sheep. Staddon quotes the early-nineteenth-century Scottish "shepherd poet" James Hogg on training the dog Sirrah by putting him in situations where his herding instincts and propensities enabled him to solve problems of increasing difficulty: "[Sirrah] would try everywhere . . . till he found out what I wanted" and in so doing would display "a great share of reasoning."

This open-ended training with eager creative learning showed its value one night when seven hundred lambs escaped from their pen and scattered onto the moors. As Hogg went looking for them in the dark, he called and whistled to Sirrah, but he searched until dawn without seeing or hearing either the lambs or the dog.

Hogg feared that all the lambs were lost. On the way home, however, he discovered that "the indefatigable Sirrah . . . looking all around for some relief, but still standing true to his charge" had gathered every single one of the lambs and herded them into a ravine. Hogg wrote, "How he had got [them] . . . all collected in the dark is beyond my comprehension. . . . If all the shepherds in the Forest had been there to have assisted him, they could not have effected it with greater propriety."

If you spend enough time around people who work with Border collies and other herding dogs, you hear about many incidents like this, in which a dog figured out a difficult new problem without human direction. I drink up these stories. They refresh and inspire my belief in the learning and reasoning abilities of dogs, and in what dogs and people can achieve when they form bonds of trust and love as fellow creatures.

David Johnson believes fervently in the reasoning ability of dogs. He told me, "Dogs and people are so much alike as far as their thinking and even their emotions go. Some people don't like that idea. I've had words with my church's Sunday school teacher a time or two, because in his teaching sometimes he would say that we are the only creatures that God made that can reason. He'd say, 'A dog can't reason.' And I'd tell him — he doesn't say it anymore, 'cause I would tell him, 'I beg to differ with you. I can give you illustration after illustration to let you know that dogs can reason, just like people can reason.' Matter of fact, as far as working livestock is concerned, I think a lot of dogs that are brought to me for training are more intelligent than the people that bring them."

As an example, he mentioned the day he sent two dogs, Maud and Gail, to retrieve some cattle from the woods beyond his pasture. Within the woods there was a fenced lot, once used for holding hogs, with gates at the north and south ends. The gates were now always left open so that cattle could move through the woods to graze in the abandoned hog lot. Here's how David explained to me what happened after he sent Maud and Gail on their mission:

I couldn't see the cattle, but that wasn't anything unusual. I could cast my dogs at livestock that was out of sight, and they would find them and bring them to me. However, this day they were gone so long that I got concerned. So I walked over to the woods. When I got there I saw Maud and Gail bringing the cows through the old hog lot. The cows went in through the north gate, but when they came out through the south gate they turned back around through the woods.

I don't know how many times they had done that, and forced Maud and Gail to round them up again. They did it once when I got in sight. I watched Maud and Gail gather them from among the trees. Their teamwork was something, because of how thick the woods were around the old hog lot. I was still far enough away, and they were so focused on the cattle, that they didn't see me watching them.

Once the dogs got the cattle together and headed back down through the north gate, I was going to intervene so the dogs' efforts didn't go to waste one more time. But something made me wait. This time, before any cows got down to the gate at the south end, Maud went around them inside the hog lot and cut off, say, eight or ten. When they went through the gate she went through beside them so she could force them to go on straight south toward the pasture.

Gail understood what Maud was doing and just stopped and waited, didn't push the other cows on through. Then Maud went back and cut off eight or ten more and brought them through the gate like that and kept them from going back north. I just stood there and watched. And that's the illustration I told the Sunday school teacher. I said, "If that dog could not have reasoned out the problem and the solution, she would have never done that."

Wayne West shared similar experiences of sending dogs to work out of sight and unsupervised. Wayne told me, "If I can't see where the dog and the stock are at, the dog has got to reason. He's got to think for himself what he has to do to get the stock from point A to point B."

Having dogs work out of sight can lead to unexpected, even un-wanted, results, but definitely ones that display independent thought on the part of the dog. David told me how he sent his dog Craig to get some sheep at the far end of a thirty-acre field in driving rain. Because of the rain and the undulations of the field, Craig could not see where the sheep were when he started out. For the same reason, David could not always see Craig as he moved across the field. David said:

All of a sudden I saw Craig start turning in too quick. I had walked up on a little rise and I could see where the sheep was at, much farther back in the field. And I thought, "What's wrong with him?" Then I see something lift up, but I can't make out what it is because of the rain.

In a few seconds I see it's a flock of wild geese flying about three or four feet off the ground. And Old Craig is right in behind them, herding them toward me. They fly like that toward me until they get close enough to see me, and then of course they lift up and go on over me.

In exhibitions I've had dogs herd ducks as well as sheep and cattle. Craig saw those geese before he saw the sheep, and he thought that's what I sent him after. I wish I had that on film, because a lot of people won't believe that it actually happened.

Wayne West and David Johnson both learned about Border collies from the grandfather of Border collies in America, Arthur Allen. The son and grandson of sheep farmers who imported top dogs from Scotland, Arthur Allen did more to foster appreciation of Border collies in this country than anyone else. He competed in and won all the important sheepdog trials. Beginning in the 1940s, he put on exhibitions with Roy Rogers's traveling show. Two of his best dogs starred in the 1955 Walt Disney film *Arizona Sheepdog*. And he wrote books that became the bibles of breeding and training Border collies in America.

Wayne told me of standing with Arthur Allen on his farm in southern Illinois and watching two of Allen's dogs disappear into a distant cornfield where the stalks were as high as a person. Half an hour later the dogs had without supervision gathered a flock of sheep out of the field and moved them several hundred yards farther away to another field through a gate that a farm employee had left open for them. Arthur Allen said to Wayne, "If I could teach the dogs to open the gate, I wouldn't have to have a hired hand."

In his book *Border Collies in America,* Arthur Allen wrote about his best champion dog, Old Tweed:

When I asked him to go get anyone in the family, he brought them regardless of what they were doing. Anyone that was on the farm very long knew when Tweed started pushing them with his nose they had no choice but to go wherever he wanted to take them. He had a wonderful ability for finding sick sheep and would come after me and would not give up until I went to see about the sheep or a ewe with a young lamb. He also had a great ability to recognize a sheep that had gotten out of a pasture and become mixed in with sheep from other pastures. He would cut him out and bring him in to be put in the pasture where he belonged. [Tweed] would do this all on his own without any help from me whatsoever. . . . He was a fine trial dog when the going was rough, and . . . when the sheep were impossible for others to handle, Tweed looked his best. But he always seemed to resent exhibition work, as he could not see the purpose behind it.

In the same book Allen says, "I like a dog that is an individualist; one who thinks for himself and will act without orders." He adds that the number one mistake that people make in training Border collies is not trusting the dogs' instincts. The number two mistake is not allowing the dogs to think for themselves. His close observation of Border collies showed how intertwined their instincts and reasoning abilities are, and how their instincts support their ability to solve problems that dogs could never encounter in the wild or could never be specifically trained to deal with in advance.

Arthur Allen's dogs Rock and Nickey (called Nick in the movie) provide beautiful examples of this in *Arizona Sheepdog,* which you can see on YouTube. One comes in a sequence that calls for Nick to herd five stray sheep for more than a day, taking an independent route from the shepherd and Rock, who are taking the main flock to their new pasture. Much of this is scripted and staged to some extent, of course, but the animals' behavior could not be scripted in every detail.

In particular there is a sequence in which the five sheep, supposedly driven by thirst, wind up in a raging river. Two of the sheep separately clamber onto a boulder and strand themselves there, refusing to move. In each instance Nick swims out to the boulder, climbs on it with the frightened sheep, and noses its feet until it falls back into the river, a behavior he could not have been taught in advance. He then herds all five sheep to safety.

Earlier in the movie Nick displays both world-class athleticism and tenderness in herding a Navajo child's pet chipmunk back into its little box. Nick's spontaneous responses to the chipmunk, including gently trapping it with his paw as part of redirecting its path, are a joy to see, an equal expression of his problem-solving skill and his compassion.

The problem-solving ability of well-trained Border collies goes along with their dedication to their work. And both these things rest on the quality of the relationships the dogs have with people. To illustrate this, David Johnson told me more about the dogs Maud and Gail:

> I was working Maud one day on some cattle, demonstrating for some visitors. She had to nip the cattle on the heels to move them, and I probably worked her for ten minutes. It was only when I called her to me with the "That will do" command that I noticed she had bit through her own top lip. One of her long top teeth was sticking through the lip. As she came out through the gate she was swiping at her face with the front paw on that side, trying to get the tooth free. Until I called her to me she never faltered, she never quit working or anything. She kept right on like it wasn't a thing wrong. She had to be in pain.
>
> Another time I was demonstrating for a visitor with Gail. She was working some sheep, and I had her bring the sheep up into the corner of a big pen. Then I gave her the command to "down." I left her there holding the sheep up in the corner while I took my visitor over to a different lot to see another dog work some cattle.
>
> We did that for probably twenty minutes, and then I put the dog that had worked the cattle back in the kennel. My visitor said, "Where is your other dog?"
>
> Understand now, we were completely out of sight of Gail and

those sheep. We walked back over there, and sure enough there she's laying right where I left her, still holding the sheep right in that corner. She was what I call a honest dog.

Some dogs are so dedicated, they want to please so strong. And then there's other dogs that aren't dedicated at all—they're just obeying you because they think they have to. Once they get an opportunity where they see you're not looking, they will start chasing the stock for fun or doing things the way they want to.

Dogs that have had opportunities as puppies to bond with people are the ones that become "honest dogs," in David's terms, and put their willpower and cognitive ability to work for their human companions with total dedication.

David's explanation of an honest dog's motivations helped me better understand the dynamics of my relationship with Chaser. Chaser's sensitive reaction to her errors during training can be extreme, and I hate to see her eyes cloud up with confusion and the hurt of disappointing me when she makes a mistake. She looks so wounded in those moments. I've sometimes worried that her strong reaction must be the result of some mistake on my part, inadvertently expressing disappointment over errors in a way that traumatized her.

David explained that this wasn't really the case: "Dogs are just like people. Some dogs are very honest, and some dogs are dishonest. I would never send a dog that's not honest to gather stock out of sight, because you can't depend on them. They get out of sight and they run the stock ragged. They do what they want to do. Some dogs just will not cheat you at all. My old Roy dog comes to mind. He was one of the most honest dogs I've ever had. He just did not want to do anything wrong, he was so dedicated. It looked like it hurt his feelings if he thought he had disappointed me."

The more I heard about Roy, Maud, or another one of David's "honest dogs," the more I understood how their honesty, as he calls it, flowed from their loving relationships with him and other people. Positive relationships with people are as crucial to a dog's development as they are to a child's. Chaser is without question an honest dog in David's terms.

Animal scientists have often worked on the assumption that animals cannot reason and are simply machines made out of flesh and blood. That view includes the idea that animals are incapable of empathy and compassion. To have empathy and compassion for another creature requires what science calls a theory of mind, an understanding on some level that other beings have independent thoughts and feelings. In recent years animal scientists have become more open to recognizing signs of empathy and compassion in animals.

Charles Darwin wrote about the social intelligence of dogs, and animal science today is increasingly recognizing the social intelligence of dogs, bonobos, crows, and dolphins, among other species. We are entering a new period of discovery about what animals' social intelligence shares with our own social intelligence, with the potential to illuminate the social component of how our own brains mysteriously acquire language. As I mentioned, researchers with backgrounds in behavioral and cognitive psychology, such as Clive Wynne and Alexandra Horowitz, and evolutionary anthropology, such as Brian Hare and Juliane Kaminski, have made great strides in delineating dogs' perceptual and cognitive capacities. They are pioneering a new understanding of dogs as thinking and feeling creatures.

The more that science discovers about the unique interspecies social relationship people and dogs share, and the kinds of learning and decision-making this relationship can engender, the more we can appreciate the wisdom and insight that farmers, breeders, and trainers have achieved. Working with dogs on a daily basis has given them a uniquely valuable perspective on dogs' capacities for problem solving and for empathy and compassion.

David Johnson said of his dog Roy:

He was a great dog to move ewes with baby lambs. Roy would move them so cautious, 'cause if he got in a big hurry and went to pushing the ewes too fast they would run over the little lambs and break a leg or whatnot. And ewes will fight to protect their lambs just like cows will fight to protect their calves. If a ewe turned and came back to fight Roy, he would nip her on the nose and turn her. But he let the little lambs wobble and bobble in against his face. He worked with his head low to the ground, and although the lambs might bump into him he would never bite them. He took his nose and just kind of pushed and rooted them along, but he never put his mouth on them. He had enough sense to know that they were fragile. Roy, if the grown ones fought him he'd nip them on the nose and turn them. But he was kind to the little ones.

David also said:

I had a customer come here one time, and he told me something that I did not believe. He said he had an old female and a young male dog. The female had gotten too old to jump into the back of his pickup truck. He said the young male dog would jump in the truck, then the female dog would rear up to put her front feet on the tailgate, and the young male dog would reach down and catch her in the scruff of the neck and pull her on up in the truck like a mother dog lifting up a puppy. He said the dogs had figured this out entirely on their own.

When the customer left, I told my dad, "That's just a story trying to impress me. The dog won't do that."

A couple of months later the customer came back. And this time he had the two dogs in the back of his pickup. He said, "I want you to watch my dogs."

He asked them to jump out, which they did. And then he told them, "Get in the truck." The young male dog jumps in. The old female rears up and puts her front feet on the tailgate. And darned if that male dog doesn't reach down and catch her in the scruff of the neck and pull her up into the truck.

I apologized to the man for doubting him, and I thanked him for bringing the dogs. What that young male dog figured out to do showed his intelligence, but it also showed his compassion. To me that's pretty amazing.

Hearing these stories from David brought to mind an experience I had in Learned, Mississippi, one July day in 1944, when I was sixteen. My sister, her husband, and I were staying with my brother-in-law's mother, Miss Lillian, on her small, hardscrabble farm. Miss Lillian may not have had much in the way of material wealth, but she was a true matriarch in her sharecropping community, respected by everyone for her wisdom, kindness, and graciousness.

We had just come back to Miss Lillian's house for lunch after a morning of Bible study in a hot, sticky classroom, and I was leaning against a maple tree debating whether to go on in the house to eat or take a swim in the nearby lake. My eyes caught something moving in the two-and-a-half-foot crawlspace under the house. Shielding my eyes from the sun with my hand, I made out Miss Lillian's old black dog, Jeb, walking back and forth in the crawlspace. He stopped for a moment to lie down, only to get back up and walk ten feet away, lie down for another moment, and then get up and move to another spot under the house.

I looked up and saw through the large picture window that Miss Lillian was preparing lunch plates at the sideboard and then passing them around to my sister, my brother-in-law, and other guests. Peering back down into the crawlspace, I saw that Jeb was following Miss Lillian's footsteps, which he could clearly distinguish from anyone else's. Jeb's devotion moved me, and for the first time in my life I pondered the significance of animal intelligence and behavior.

The devotion of Border collies to a beloved master is so well documented that it cannot be called merely anecdotal evidence. David imported his dog Roy, then three years old, from northern England, where a shepherd had raised and trained him from the time he was weaned from his mother. David told me, "There are so many instances where a shepherd has been out with a dog, and maybe because of extreme weather conditions, an accident, or some illness the shepherd falls unconscious or dies. And when that shepherd is found, he may be covered with three feet of snow, but the dog would be laying right there with him. After I got Roy, he laid back in the kennel for three weeks and would not work at all. He would eat and drink, but he wouldn't work. He had to finish grieving for his old master before he could accept the idea of working for me."

A dog can have trouble adjusting to a new master for practical reasons, as well. In the late 1990s Wayne West imported a dog named Ben from one of the most respected breeder-trainers in Scotland. However, Ben just would not work for Wayne. Despite all his experience, Wayne was flummoxed. The man who sold Ben didn't know what the problem could be either. Eventually he and Wayne guessed that it might be a question of accent. Several days later, Wayne received a cassette tape on which Ben's previous owner had recorded the standard Border collie commands in his thick Scottish brogue. Wayne told me, "You know I can't carry a tune, so I wasn't sure how I'd do. But I listened to that tape and imitated it for hours. When I took the dog out and worked him after that, he fulfilled every command faster than you can say Jack Robinson."

I got a chance to witness Wayne working with Ben. His South Carolina twang imitating the Scottish brogue made everyone but Ben laugh. Wayne didn't sound remotely like a Scotsman, but he reproduced the inflections well enough so that Ben finally understood him. Ben was as happy about that as Wayne. He had been bored stiff from inactivity. He wanted nothing more than to work sheep and fulfill himself as a Border collie. I was happy about Wayne's finally being able to bond with Ben at the time, and even happier a few years later when we got Chaser, because Ben was her daddy.

Along the same lines, David said, "Probably the best female Border collie I ever had was Pat. If I was talking to some people, Pat came and sat down by me. She looked from one person to another as they spoke in turn. She focused so intently on the person who was speaking and listened so attentively to whatever they were saying that it was just like she was part of the conversation."

The Border collie's sensitivity to human speech was one of many traits that I expected Chaser to display. But as I've mentioned, I wasn't prepared for the extent of that sensitivity. Seeing Chaser's reaction to variations in tone of voice made me much more aware of the tone I adopted, not just in speaking to her, but in all my conversations with Sally and others.

Border collies' sensitivity to the human voice complements one of their most remarkable instinctual gifts, the "eye." Uniquely among dogs, Border collies often control and move livestock simply through the power of eye contact — their "Svengali gaze," as an August 31, 1953, *Time* magazine article on Arthur Allen put it. A Border collie's "eye" on the stock can be loose, overbearing, or a happy medium. The "eye" is genetically constrained, and this trait fades out when Border collies are crossbred with other breeds.

Wayne West first told me about the Border collie's ability to give "an eye to the sheep and an ear to the farmer." The best dogs can win a battle of wills with unruly stock just by "putting that stare on them," in David Johnson's words. I saw this when Wayne worked his own dogs or put on exhibitions with David and other Border collie people. Arthur Allen wrote that his dog Nickey "had the strongest glare in his eyes ... [and] was the most fearless and forceful without grip of any dog I have ever worked." The same dog who was so tender with a child's pet chipmunk could even herd a mountain lion with his "Svengali gaze." Disney's *Arizona Sheepdog* features scenes with a mountain lion. Both Wayne and David told me how during filming, the mountain lion's trainer relied on Nick to herd the big cat into his cage.

Like her extreme sensitivity to tone of voice, Chaser's own "eye" was something I had to get used to. Early on I often thought that she wasn't paying any attention to me at all when I spoke to her about something I wanted her to do. As soon as I started to give her an instruction, she lay down and never looked directly at me. I had to remind myself that Arthur Allen said, "I like a dog that will listen for orders and give me his ear and only the slightest glance of his eye."

Today when Chaser and I do exhibitions, people regularly comment that she seems to ignore everything but my voice. In blind experiments, as they're called, when Chaser can't see me or another person who might be giving her instructions, she does indeed respond to voice alone. But on other occasions, she sees me and my gestures with a slight glance of her peripheral vision. She aims her strong "eye" on the balls and other toys that have become her surrogate sheep.

The teaching wisdom and field reports of those who spend their lives working with Border collies are resources that animal science needs to appreciate more fully. Above all, researchers need to pay heed, as John Staddon urged, to the way great Border collie trainers recruit the power of instinct to foster a dog's creative problem-solving ability. In that spirit let me close this chapter with a saying of Arthur Allen's that I have found valuable in all my interactions with Chaser: "Try to make it a fifty-fifty proposition of you trying to understand your dog while your dog is trying to understand you."

8

Learning by Playing

"C HASER! THIS IS BLUE."

On my knees, I roll the blue racquetball toward her on the living room floor with no other toys anywhere in sight. As Chaser takes the ball in her mouth I repeat "Blue," the name of the ball. She begins to chew on it as I crawl toward her. When I'm close enough to reach out and grab the ball, she runs away, then turns and looks at me, wagging her tail and grinning.

In Chaser's first two months with Sally and me, she's demonstrated that she loves racquetballs, which happen to be blue in color, because they are so light and bouncy and feel so good in her mouth. And if I'm chasing her to get a racquetball, it must be very valuable. That makes playing keep-away with it exciting, fun, and, I hope, memorable.

"Blue," I whisper as I again reach for the ball. She darts just out of range again. Turning to look at me and wagging her tail in delight, she drops the ball accidentally and chases its bounces across the room.

Each time she takes the ball in her mouth I repeat its name: "Blue. That's Blue, girl, Blue." The ball bounces where I can grab it, and I am determined to teach her the name of the ball as I roll it to her, repeating, "Blue. Blue. Chaser, that's Blue!"

There was no road map in the scientific literature for teaching words and their meanings to a dog or any other animal. I began trying to teach Chaser words on the assumption that unless the words had strong positive value in her mind, she would not be motivated to focus on them and remember them. I reasoned that the best way to give words positive value was to associate them with objects used in play. Play would give the objects value, and by extension play would give value to the names of the objects and the verbs and other words that directed play activities.

Play should have many learning benefits, I knew. As a college professor, I had seen how play frees the mind from tension and anxiety, thereby opening the door to creative thinking.

I wanted to start language-related play with Chaser as soon as possible. Research has shown that there is a critical early developmental period for children to acquire language when they are toddlers, and I speculated that dogs might have a similar developmental window as puppies.

My hunch was based on my experience, shortly after I came to Wofford, as a consultant to the Spartanburg public schools. I administered hundreds of tests individually to sixth-, seventh-, and eighth-graders to assess their cognitive development and need for special education. In many cases it seemed plain that a student's learning problems stemmed from a lack of stimulation and nurturing in early childhood. It was especially sad to see the number of children with such problems, because there was little anyone could do to ameliorate things at that point. The critical time to help these children is in the earlier years.

The bottom line was that I didn't know if there was a communication-related developmental period for dogs as puppies. But if there was one, I wanted to make the most of it.

In anticipation of getting a puppy, I went to the local thrift store and bought a shopping cart full of secondhand children's and pet's toys: balls, stuffed animals, rubber animals and dolls, leather chew toys, pull toys, Frisbees, and so on. I was going to use these toys to teach Chaser proper nouns, words that uniquely name something or someone. Because of their strictly limited meaning and one-to-one association with something or someone, proper nouns, such as Mama, Dada, and their own names, were the first words toddlers learned. So I would begin teaching Chaser words with proper nouns, too.

I gave each object a proper noun name — Elephant, Lion, Santa

Claus, and so on—and wrote the name on it in permanent ink. The only exceptions to my naming of toys were blue racquetballs and green tennis balls. Any racquetball was "Blue," and any tennis ball was "Tennis."

In the first two months after we got Chaser, I played with her with all of these toys. But I introduced each toy by name—"Chaser! This is Blue!"—one at a time, with no other objects on the floor available for play. Chaser responded enthusiastically to each new addition to her flock of playthings. She treated the objects like surrogate sheep and eagerly chased, stalked, fetched, and gathered them. She quickly learned the obedience and herding commands in the course of this play. Almost immediately she also started to show flashes of getting the name of an object into her short-term memory. Those moments really excited me, especially as they began to come more frequently and the memories began to be more long lasting, extending from one play session to another a few hours later, and soon a day or more later.

I wasn't yet seriously concentrating on teaching Chaser the proper noun names of the toys. Teaching her obedience behaviors took precedence in order to keep her safe from traffic and reliably direct her movements during training. I also wanted to build up the play value of the toys in her mind. And in that respect I was observing how she most liked to play with the toys and which ones she favored in different circumstances. With that information I could heighten her interest in a toy, and thus in its name, when we began training intensively.

During this time I also established a foundation for language training games by teaching Chaser verbs such as "fetch" and "find," in addition to those for the obedience and herding behaviors. I taught Chaser the verbs for actions she produced as part of her natural repertoire by naming the actions when I saw her do them. For example, Chaser often shook a toy after she picked it up in her mouth, and so I repeated "Shake it" several times whenever I saw her doing this. The behavior of shaking a toy eventually came under stimulus control, and Chaser shook a toy whenever she heard "Shake it." Among other verbs I taught her in this way were "drink," "run," and "bark."

For behaviors that she spontaneously displayed less frequently, I first created a context that evoked or enhanced the likelihood of the behavior. For example, I wanted to attach "find" to Chaser's natural tracking behavior. To do this I let Chaser smell an object with a strong scent, and then hid the object out of sight. While Chaser followed the scent trail, I repeated "Find" out loud. In a few instances I used food as a lure to guide Chaser's behavior to a desired response, such as turning around in response to "Turn around."

Later on there were more complex behaviors I wanted Chaser to produce on verbal command, and teaching these required shaping by successive approximations. But we only needed a handful of simple action verbs to get Chaser's proper noun learning under way.

Chaser had been with us only three weeks — we were still calling her Puppy — when Alliston Reid e-mailed me: "I just ran across this new article and know you'll want to see it. Talk to you soon." Alliston's e-mail included a link to a paper that was a few weeks younger than Chaser. The June 11, 2004, issue of *Science*, the most influential scientific journal in the world along with the British journal *Nature*, included an article called "Word Learning in a Domestic Dog: Evidence for 'Fast Mapping,'" by Juliane Kaminski, Josep Call, and Julia Fischer of the Max Planck Institute for Evolutionary Anthropology in Leipzig, Germany. The paper reported that a nine-year-old Border collie named Rico knew more than two hundred proper noun names that his owners had given to various toys.

This was far beyond what previous studies had found. The finding also merited attention because Germany's Max Planck Institutes are the equivalent of America's national laboratories and National Institutes of Health. The paper's lead author, Juliane Kaminski, headed up a research group at the Max Planck Institute on the evolutionary roots of human social interaction. Kaminski and her group searched for clues to the origins of human cognitive capacities and social behaviors in studies of bonobos and other animals, and she had established an international reputation in animal science. As Alliston said to me when we caught up on the phone, "These are heavy hitters."

Despite the assertions of trainers and farmers that Border collies, as well as other herding and working dogs, could learn a sizable number of words, prior studies had found that dogs knew only around a dozen words, most of them verbs like "sit and "stay." In contrast, the article said, "Rico's 'vocabulary size' is comparable to that of language-trained apes, dolphins, sea lions, and parrots."

Moreover, the article said that Rico was capable of learning by exclusion, or "fast mapping" as it has been dubbed in studies of young children. This meant Rico could fetch an unfamiliar object from a group of familiar ones after hearing its name for the first time.

Summing up their findings, the paper's three authors observed that they could not say if Rico's abilities were the result of "an exceptional mind" or his training. Noting that "dogs appear to have been evolutionarily selected for attending to the communicative intentions of humans," they wondered if "some of [Rico's] talent may be accounted for by the fact that Border collies are working dogs" who follow verbal commands in herding livestock. They asserted, however, "Our results strongly support the view that a seemingly complex linguistic skill previously described only in human children may be mediated by simpler cognitive building blocks that are also present in another species."

When an article counters the prevailing scientific consensus or addresses a long-standing debate on a major topic, *Science* sometimes publishes an accompanying "Perspective" on it written by a leading scholar. That was the case here, with a "Perspective" titled "Can a Dog Learn a Word?" by the childhood language learning expert Paul Bloom of Yale University. Bloom accepted that Rico could correctly identify more than two hundred objects by their proper noun names by fetching them on command. He even said, "For psychologists [studying language learning], dogs may be the new chimpanzees." But he expressed strong reservations about whether Rico really understood that each name had an independent meaning apart from fetching the object associated with it.

Rico's owners said that on command he could also perform other actions with the objects whose names he knew. But the Max Planck Institute researchers had only tested Rico in fetching, and Bloom rightly insisted that that was the only evidence science could evaluate. Among other things, Rico's owners might sincerely believe their dog knew how to do these things but not realize that they were unconsciously cuing him as to what to do.

Bloom noted that human toddlers begin to learn words by associating them on a one-to-one basis with specific things, individuals, and actions. But toddlers quickly develop and display the understanding that words can be used in different ways in different contexts. In addition to learning that proper nouns refer to specific, unique things, toddlers also quickly learn that common nouns refer to whole categories of things. And they can learn words simply by overhearing them, whereas "Rico . . . learn[ed] only through a specific fetching game." He concluded, "Rico's abilities are fascinating, but . . . it is too early to give up on the idea that babies learn words and dogs do not."

The Rico paper and Paul Bloom's critique of it provided several goals for my research with Chaser. One striking fact was that the most successful language-trained animals, no matter what their species, all had roughly comparable vocabulary sizes of around one hundred to five hundred words. When the results of a number of experiments cluster, you have to ask if that apparent pattern is the whole story. I wondered if one hundred to five hundred words was the ballpark all animals were in, or if more was possible, and I wanted to test the limit.

I also wanted to see if Chaser could make the leap that toddlers make and understand a variety of words — proper nouns, common nouns, verbs, and modifiers — according to how they are used in different contexts. Exactly what makes this leap possible in children is unknown. But there is no doubt that it happens as every child acquires language. There is also no doubt that one-to-one associative word learning is a necessary first stage along the way. For Chaser to understand that some words represent categories and to learn concepts, our best chance was building a cycle of positive reinforcement in associative word learning.

The Rico paper did not explain how Rico's owners trained him to learn an object's name. But given Chaser's fast-developing memory for the names of her toys, the appearance of the article at just this time felt like a sign that we were on the right path.

As soon as the Jeep incident was behind us, I began to do more intensive training on the names of objects. The procedure began with Chaser sitting in front of me in the living room. So that Chaser could learn without errors and build on a series of small successes, the only toy in sight was a little stuffed Santa Claus doll that we'd played with many times, and whose name — Santa Claus — I had said dozens of times over the previous two months.

Holding up the doll and pointing to it, I said, "Chaser! This is Santa Claus."

Saying "Pop-Pop hide Santa Claus," I lowered the doll to the floor in plain view. Chaser's gaze never left it, and I trusted that she was giving me her ear and a glance of her eye.

I dropped the doll on the floor and said, "Chaser, find Santa Claus," as she darted forward to pick it up in her mouth.

"Good dog!" I told her. She looked up at me triumphantly, tail wagging, ears pricked up, and eyes bright and wide, eager to find out what was next.

Stepping back a few feet I said, "To Pop-Pop, to Pop-Pop. Here," beckoning her to me with outstretched arms. She came to me and let me take the doll from her mouth, as I again said, "Good dog!"

Tossing the doll to her in a high arc, I said, "Chaser! Catch Santa Claus. Catch Santa Claus." She reared up on her hind legs to catch the doll, and as her front feet hit the ground again her tail went up and wagged back and forth with excitement and pleasure. "Good dog, good girl!" I said, positively reinforcing her.

"To Pop-Pop. Here," I said. She brought me Santa Claus, and her tail wagged faster at hearing another enthusiastic "Good dog!"

Finally I tossed Santa Claus into the middle of the living room and put Chaser through the herding behaviors: "Chaser, come by. Come by Santa Claus. Whoa. Way to me, way to me. There, there. Drop. Drop behind Santa Claus. Chaser, crawl to Santa Claus. There. Chaser, one, two, three, take Santa Claus! Good dog!"

After this herding play with as many repetitions of the name Santa Claus and as much positive reinforcement as possible, we began all over. To keep the game interesting I progressively hid Santa Claus so that the doll was harder and harder to find. In each word training session, consisting of a brief trial at finding Santa Claus and a few minutes of play with it, I said the name twenty to forty times. Over the course of the day's training sessions, we went through this exercise with Santa Claus at least twenty times.

Over the next few days we continued to play this game with Santa

Claus. We also played the game with other objects that were already familiar to Chaser. The excitement of finding the object, whether it was hidden in plain view or in the other room, put a grin on both our faces and got Chaser's tail wagging with pleasure. After a brief moment to celebrate that victory, I rewarded Chaser with play with the object for three to five minutes. Depending on the object, the play might involve a little tug-of-war, chasing her as she scampered away with it, throwing it for her to catch or fetch, and most of all herding games. All the while, I repeatedly said the name of the object in very simple statements that described what we were doing and positively reinforced her with praise and pets: "Chaser, catch ___. Good dog! Chaser, shake ___. Good dog! Chaser, fetch ___. Good girl, Chaser!"

After a few trials with different objects, Chaser knew what to expect, and her behavior briefly made me fear she had lost interest in what I was doing. When I said, "Chaser, this is ___," she lay down on the floor and tilted her head as if she were completely ignoring the object and what I did with it. She remained lying on the floor, seeming utterly complacent, even bored. But as soon as I said, "Chaser, find ___," she sprang to her feet and raced to find the object.

As she did so I repeated, "Chaser, find ___, Chaser, find ___," in a soft, encouraging tone until she found it. This I greeted with a triumphant "Good girl, Chaser!" followed by a round of play with the object.

We did this many times for each toy, and I was curious to see whether as time went on Chaser needed fewer repetitions to learn the name of a toy. My criterion for saying she had indeed learned a word had two stages. First she had to select the correct object out of a group of objects that included seven other objects whose names she had learned. I called this the 1-of-8 test. And then she had to do the same thing seven more times in a row over the course of a day or so, each time with a different set of seven other objects whose names she knew as distracters. I called this the 8-of-8 test. After asking for the newly learned object in a group of eight toys, I also asked for the seven previously learned objects by name as a way to rehearse and test prior learning.

If Chaser missed fetching any object by its name in the 1-of-8 or 8-of-8 tests, I gave her more training and play with that object alone, with no other objects in view. She missed here and there, but her performances on the 1-of-8 and 8-of-8 tests were usually 90 to 100 percent correct.

As we did this with more objects I also needed to assess Chaser's long-term memory of the objects' names. I had to be sure that Chaser wasn't freeing up memory space for new object names by forgetting older ones. So every month I informally tested Chaser's retention of her complete vocabulary, to that point in time, with random sets of twenty objects on the floor in front of her. In all these tests over a period of three years, Chaser always got at least eighteen of twenty right. At the end of that time it took fifty-one sets of twenty to test her long-term retention of the names of 1,022 toys, a process that I spread over about two weeks every month.

I became an expert at evaluating toys for Chaser's play and training. I could take a brief look at a stuffed animal and gauge if it would last a week or a year without needing to be sewn up. I was always looking for the sturdiest ones possible. The growth in Chaser's flock meant that there were multiple Frisbees, balls, and stuffed lions, tigers, and bears. Despite these similarities, every object had to have unique features that Chaser could distinguish visually. Over time it became harder and harder to find toys that weren't highly similar to ones we already had.

Occasionally Sally and I splurged ten dollars on a dog toy from the pet store, especially during after-Christmas sales. And I was always looking for the perfect Frisbee. I tried to stick to cloth Frisbees as opposed to plastic ones with hard edges. But I bought the majority of the toys mainly at the Salvation Army store, which was closer, and the Miracle Hill thrift shop, which was much cheaper. The Salvation Army store wanted seventy-five cents or a dollar per toy, whereas Miracle Hill would sell a big plastic garbage bag full for a couple of bucks.

Sally was extremely patient as the house filled up with toys. Although we bought many large Rubbermaid containers to store them, there were always at least fifty toys strewn around for training, testing, and pure play. But sometimes the clutter of toys got to be a little too much for Sally and we had to do a rapid cleanup. This got a little easier and more pleasant once Sally decided to teach Chaser to clean up her own toys the same way she'd taught Robin and Debbie when they were small and later helped Debbie teach Aidan.

"Clean up, clean up. Everybody clean up," Sally sang, as she picked up one of Chaser's toys and dropped it into a plastic storage tub in the middle of the living room. Attracted by her singing, Chaser looked at Sally, head cocked to one side with curiosity. "Clean up, clean up. Everybody clean up," Sally continued singing. She picked up two more toys and gave one to Chaser, who took it eagerly in her mouth. Continuing to sing, Sally pointed at the plastic tub and dropped the toy she was holding into it. Chaser brought her toy over in response to Sally's song and pointing, and dropped it into the plastic tub.

"Good girl, Chaser!" Sally said, before resuming her song and picking up more toys for Chaser to take and drop in the plastic tub. Midway through the cleanup, Sally stopped picking up toys for Chaser and simply pointed at the ones that were still scattered around on the floor and the furniture. It didn't take too many repetitions of this song and coaching for Chaser to learn the cleanup game. She enjoys it, too, coming eagerly when she hears someone sing, "Clean up, clean up. Everybody clean up."

Later I learned that telling Chaser "Clean up, and we will play Frisbee in the yard" makes her even happier to play the cleanup game. She darts around picking up her toys, and as soon she has dropped the last one into a Rubbermaid container she heads for the door with her tail wagging.

Sally also sewed up dozens of toys over the years, especially when she became exasperated by my duct tape repairs. That did not include beanbag-type toys, however. Early on we learned to avoid these after Chaser shook one vigorously, a seam ripped, and the beans flew everywhere.

As Chaser's flock expanded to include more named objects, the likelihood that her correct answers were the result of chance became smaller and smaller. That was because the possible answers were not only the named objects in the groups of eight or twenty, but all the objects whose names she had learned at the time of the test. The more object names she learned, the more impressive her consistent, very high level of accuracy became.

The probability of her correct answers resulting from chance was small in her first tests, and her correct answers were well above the threshold that made them statistically significant. In the later tests the probability that her correct answers resulted from chance was vanishingly small. Chaser's proper noun learning reached the point where the probability that her correct answers resulted from chance was less than 1 in approximately 20 million.

However, the informal 1-of-8, 8-of-8, and 20-of-20 tests were subject to criticism that I was, knowingly or unknowingly, giving Chaser visual cues about which object to fetch. So every month I did a formal 20-of-20 test as well. I took twenty objects from her flock at random, reaching blindly into the plastic tubs where we kept them, and wrote their names down. After brief rehearsals with each of these twenty objects, I put them in another room so that neither Chaser nor I could see them and I could not possibly give her an unintentional visual cue about which one I wanted. And then I asked Chaser to go and fetch each one by name.

The first time we did this, Chaser got 18 of 20 right. Thereafter her performance on the formal 20-of-20 tests was consistently in the 90 to 100 percent range. In fact, she always got at least 18 out of 20 correct on the formal as well as the informal test. When she missed fetching an object by name in any 1-of-8, 8-of-8, or 20-of-20 test, I gave her additional training with that object alone, with no other toys in sight, before testing her again.

All told, our language training and testing sessions amounted to four to five hours a day. Weather permitting, I gave Chaser two to three hours of other physical activity a day outside the house. Some of the outdoor time was devoted to vigorous play with her named toys. But it also included agility play, tracking and stalking, and hikes in the nearby mountains.

In the backyard I set up an agility course with jumps, a tunnel, obstacles to crawl under, and stakes to weave in and out of. Chaser loved running the course, and my directions for her to run it in different directions and patterns were another opportunity to teach her words. Every run through the course ended with my sailing a Frisbee for her to catch triumphantly.

Some days we took the forty-minute drive to sixty acres of forest we own on the Tyger River, southwest of Spartanburg. Chaser loved jumping over logs and exploring under bushes as we hiked to a spot on the river with a wide bank of flat rock. This was perfect wading water for her to splash into while retrieving sticks and balls I tossed for her. Walking the land, she darted here and there to investigate animal scents, always on the move, until we climbed back into the car, where she curled up and snoozed until we got home, at which point she was ready for round two.

At five months old Chaser knew more than fifty words, including those for the basic obedience behaviors. More important than the number was that we were both having lots of fun and she was eager to herd more toys, and more names, into her growing flock. Outside of training sessions, she often initiated play with her named toys and loved engaging in activities with them that mimicked herding sheep. A sequence of chasing, catching, bringing back, and gathering together a group of toys was deeply involving and rewarding for her.

As Chaser approached six months of age, intensive practice was speeding up her word learning so that she needed fewer and fewer trials per word. I could see her response time — animal scientists call it latency — getting shorter and shorter between the moment when I said, "Chaser, find ___ " and the moment when, eyes shining and tail wagging triumphantly, she brought back the correct surrogate sheep. She seemed to be learning words so quickly that I decided to try an extreme test.

I brought out a new object, a fleece-covered brown and white stuffed pony, about nine inches high and about fourteen inches from nose to tail. On the pony I had written the name Puddin. With Chaser sitting in front of me, I pointed to the stuffed pony and said, "Chaser! This is Puddin." Then I immediately took it into the bedroom and placed it on the floor among seven other objects. Four were familiar objects whose names Chaser already knew, and three were novel objects she had never seen. Including three completely unfamiliar objects in addition to Puddin made the test very stringent, because it dramatically increased the possibility of error.

I came back in front of Chaser and asked her to find two of the previously learned objects, which she did without a hitch. There were still six objects on the floor in the other room: two familiar ones whose names she'd learned previously through repeated trials; three completely novel objects; and one, Puddin, that she'd seen and heard the name of only once. I then said, "Chaser, find Puddin."

Chaser sprang to her feet and dashed into the next room. In another flash of fur she stood before me wagging her tail and grinning from ear to ear with Puddin in her mouth. I immediately tested her twice more, and four more times the next two days. Each time I put Puddin down among a different set of seven other toys. Each time Chaser was perfect.

There was no doubt about it. Chaser had learned Puddin the pony's name in a single trial. Identifying the new object correctly after hearing its name only once indicated that Chaser had achieved a form of referential understanding. Somehow she had grasped the idea that objects can have names. She had learned that my pointing to an object and saying "This is ___" meant I was going to announce the object's unique name. To use the terms of childhood language learning research, Chaser had learned two referential social cues for indicating word meaning. On top of associative learning she had now added at least the first stage of intuitive learning via symbols. This was supposed to be impossible for nonhuman animals.

Chaser's intuitively understanding that objects can have names was a defining moment for her as a learner. Chaser did not consciously realize what had happened, any more than eighteen-to-twenty-four-month-old toddlers consciously realize that they are suddenly understanding words in a new and fuller way. But I knew that Chaser had crossed a threshold and entered a whole new world of learning.

Random testing showed that Chaser could learn as many as ten new words a day, about as many as a nine-year-old child learns. Unlike a nine-year-old child, however, Chaser needed very extensive rehearsal time to retain this many words. So we settled into a pattern of one to two new words a day, which she could lock into long-term memory with a few days' worth of rehearsal sessions off and on through the day. Brief pilot testing suggested she could have learned three to four proper nouns a day, but I was also focusing on teaching her other elements of language, including common nouns and learning by exclusion.

At the age of seven and a half months, Chaser knew more than two hundred words, as many as Rico. That Chaser was now reading my mind through the words I spoke to her gave me goose bumps. It made me think of the moment when seven-year-old Helen Keller first understood the connection between words and the things they represent. For more than five years after a terrible fever left then nineteen-month-old Helen blind and deaf, she lived in darkness. She was almost like a feral child, subject to violent tantrums of frustration. Her desperate parents brought in twenty-year-old Anne Sullivan, herself nearly blind, to try to teach Helen. Every day Anne traced letters on the palm of Helen's hand, trying to communicate that these tracings spelled words that represented objects such as a doll she gave the little girl as a get-acquainted present.

Helen could not grasp that there was a word for everything and that everything had a word — or many words — to describe and represent it. In a rage at not being to able to understand what Anne meant by signing the letters *m-u-g* in her palm, Helen broke the doll that she was carrying everywhere.

About a month later, Helen and Anne were at a water pump on the Kellers' farm. As water poured over Helen's hand, Anne spelled out *w-a-t-e-r* on the palm of her other hand. Over and over again Anne spelled the letters: *w-a-t-e-r, w-a-t-e-r, w-a-t-e-r, w-a-t-e-r.*

Suddenly Helen burst out with a cry of joy. It was the defining moment of her life. The eyes in her mind opened and she saw that *w-a-t-e-r* meant the substance pouring over her hand, the substance she drank and encountered in other ways every day. Her genius at last set free, she was now on the path to gaining full command of language. Her learning exploded, and she grew up to become one of the most extraordinary and influential figures of the twentieth century.

Seeing Chaser's vocabulary increase day by day and week by week, I raised my sights from merely matching Rico to surpassing him. The goal for learning proper nouns was now a thousand words. A thousand-word vocabulary would be enough to show that Chaser's long-term memory system was extensive and robust. It was also enough to demonstrate her understanding of words as more than object names and in more contexts than simply fetching objects. If the skeptics didn't find Chaser's knowledge of a thousand proper nouns in combination with multiple common nouns and verbs convincing, they weren't going to be more favorably impressed by two thousand or three thousand proper nouns.

In the meantime I wanted to show Wayne West what a smart puppy we'd gotten from him, and I wanted to talk to him about giving Chaser a chance to herd sheep.

I put about twenty of Chaser's toys in the back of my pickup truck and drove out with her to Wayne's place late on a mild fall afternoon. It was Chaser's first return to the place of her birth. She was now almost her full adult size, about twenty inches high at the shoulder, but she hadn't filled out to her full weight and strength yet. She was still very much a puppy. I saw no sign that she recognized the surroundings or Wayne. But true to her social nature, she was delighted to get his warm welcome.

Wayne's backyard of nicely mown grass was separated by chainlink fencing from his kennel and barn area and the beginning of his pastures. Five sheep were in a large pen, about one quarter the size of a football field, next to his kennel. But Chaser paid no attention to the sheep or the dogs in the kennel as I got the plastic tub of her toys out of the back of my pickup and dumped them onto the grass. The appearance of her toys meant fun, and she kept her eye on her surrogate sheep.

Wayne watched in grave silence while Chaser retrieved one object after another by name and I directed her in herding-like play with them. Turning to Wayne, I said, "That's just a fraction of her learning. Right now she knows about two hundred words, and I'm thinking she can get to a thousand or more."

Wayne cracked a smile and said, "Doc Pilley, you've got a lot of patience." He wasn't at all surprised that a Border collie could learn a couple hundred words, and he didn't consider it a stretch for a Border collie to keep several hundred items straight in his or her mind. Working Border collies might have to distinguish individuals among hundreds of sheep from different flocks, and mingle or separate the flocks on command. He told me, "With you as a teacher, she'll hit the thousand mark and beyond. I don't have one iota of doubt about that. You're a dog man for sure, Doc."

That meant a lot to me and emboldened me to ask, "You've got those five sheep over in the pen there. Do you think we could try Chaser on them? Do you remember you let me try that with Yasha when he was about a year old?"

"Vaguely." Wayne smiled and added, "Now that you mention it, I seem to recall that Yasha had plenty of herding drive."

Not having been born on a farm, Yasha had never even seen a sheep when I took him out to Wayne's. And I hadn't taught him the herding commands, as I had recently taught Chaser. When Wayne and I walked into the sheep pasture behind his house with Yasha, I excitedly said, "Yasha! Fetch the sheep."

Yasha instantly took a wide arc out around the sheep. I said to myself, "By gosh, he has the instinct." Yasha then rushed toward the sheep, and they bolted for the barn. A lamb couldn't keep pace, however, and Yasha blocked his path, ignoring the other sheep. I was about twenty-five yards away from the lamb in the opposite direction, and the lamb raced to me and pressed his body against me for safety. Yasha came up behind him and spontaneously dropped to the ground, in good Border collie fashion, about ten feet away. He was a natural, I thought.

Wayne laughed on being reminded of the details. "Chaser's young yet, Doc," he said. "But I've seen dogs as young as twelve weeks old move sheep. And Arthur Allen started training his movie dog Nickey when he was four months old, because the dog was so darned precocious. Let's give it a try. But we both got to be ready to come to Chaser's aid. You've got to back a young dog up sometimes so they don't lose confidence."

I put Chaser's toys back in the plastic tub and stowed the tub in the pickup. Wayne was waiting for Chaser and me at the gate to the sheep pen. Even as we approached the gate, Chaser showed no particular interest in the sheep, who were grazing in a loose cluster twenty or thirty feet from the gate. She'd enjoyed our play with the toys and she walked at my side, tail up and wagging slightly, ready for some more fun, whatever that might be.

We all went through the big swinging gate. Chaser and I remained just inside the gate while Wayne walked into the field at an angle until he was roughly equidistant from us and the sheep. I looked down at Chaser.

Aha! She was eyeing the sheep intently now.

"Chaser, go out," I said.

She left my side and trotted out into the field toward the sheep, but kept twelve or fifteen feet distance from them, just as a Border collie should. When she was even with the sheep I said, "Chaser, come by," and she circled around behind them. When she was directly opposite me on the other side of them, she stopped and dropped to the ground on her belly on my "There" and "Drop" commands. I asked her to "walk on" toward the sheep, and she did so, slowly but surely. When she was seven or eight feet from the sheep, four of them began to move away from her. But the fifth sheep, a large ewe, didn't budge, and the rest of the flock stopped moving and remained near her. Chaser kept coming forward, but her steps slowed.

Sheep get to know dogs, just as dogs get to know sheep. The big ewe had been herded by many dogs over the years, as Wayne worked his experienced dogs for visitors and customers and trained his young ones.

Chaser was only four feet away from the sheep. She was still moving in toward them, but only one careful step at a time. *Should I call her off?* I wondered. Her ears and tail were up, and she was focused on the sheep. I glanced over at Wayne. He was watching carefully but did not look concerned. So I said nothing.

The next thing Wayne and I knew, the big ewe charged out of the little flock straight for my puppy!

Chaser darted out of her path but wasn't quite quick enough. The ewe butted Chaser on the hindquarters and rolled her onto her side.

Wayne and I both rushed up to get between Chaser and the ewe. Chaser was already fifteen feet away, in no danger. And in no apparent distress, either. She didn't have her tail down between her legs in fear or anxiety. That was a relief to see, and I told her she was a good girl and had done well. But Wayne and I quickly took her out of the pen, leaving the sheep almost exactly where they were when we'd entered it a few short minutes before.

Although Chaser seemed her normal self, I didn't want being butted and rolled down onto the ground by the sheep to be her last experience of this visit to Wayne's. I got a Frisbee out of the pickup and we played with it for a few minutes. While we did that I asked Wayne, "Is this experience going to ruin her for working sheep?"

"That's unlikely, Doc. She's happy as a clam now. She didn't get her confidence hurt serious."

"What's the next step, besides waiting till she's older?"

"Well, you do gotta wait until she's at least a year old, probably. And then we can put her in the pen with some sheep and one of my experienced dogs. I'll work my dog on the sheep and Chaser can watch and learn, and she'll probably just naturally start imitating what my dog does."

"That sounds great, Wayne. Thank you."

"My pleasure, Doc."

"I was too impatient," I said.

"Don't worry yourself none about that, Doc. Chaser'll be fine," Wayne said. He grinned at me and added, "You just carry her on home and knock yourself out teaching her more words."

Driving home with Chaser curled up on the passenger seat, I couldn't stop blaming myself for my impatience. But I was happy about how Chaser had behaved all the way through. There was no doubt that she had the instinct and desire to work sheep, and the little scuffle with the ornery ewe hadn't made her downcast or frightened. I also chuckled to myself over Wayne's reactions. He wasn't surprised by Chaser's learning. He knew Border collies were plenty smart. But behind the wisecracks about my teaching, I knew he was happy to see what she could do and would be eager to hear about her progress.

I was sure Wayne was right that Chaser could learn a thousand words. I hoped he was also right about her being able to herd sheep when she was a little older.

9

Herding Words

TEN-MONTH-OLD CHASER was restless. She brought Merlin, a little wizard doll, over beside my chair and shook it, then bowed her front legs and wagged her tail, eager for play.

"No, Chaser. Not now, girl," I said. It was a Sunday evening in late February and Sally and I were just starting to watch the PBS series *Nature,* one of our favorites. Tonight's documentary was entitled "Snowflake: The White Gorilla." I was immediately caught up in the story of history's only known albino gorilla, who was born in Equatorial Guinea, captured by villagers because of his white fur and pink skin (the result of inheriting a recessive gene for albinism from both parents), and then, thanks to the intervention of a Spanish primatologist, taken to live in the Barcelona Zoo.

Fifteen minutes later I saw Chaser out of the corner of my eye and turned to look at her. She stood behind seven of her toys: Merlin, Uncle Fuzz (a stuffed monkey), Choo Choo (a rubber squeak toy in the shape of a railroad locomotive), Cinderella (a fuzzy pink poodle in a blue ballerina tutu), ABC (a cloth cube with those letters on its faces), Stubborn (a stuffed mule), and Santa Claus.

She'd learned some of their names early on and others only recently. At this point she knew more than three hundred objects by their proper noun names, and we were continuing to build her vocabulary by one to two new object names a day. The formal tests of long-term memory I did every month now included a hundred objects in five 20-of-20 trials, where Chaser had to retrieve each of twenty objects by name when they were all out of sight in another room.

Chaser had put the toys on the floor between us like sheep gathered together between a working Border collie and a farmer.

"No, Chaser," I said.

"No" had long since lost any negative emotional quality for Chaser. Depending on the tone of voice in which someone said it, the word meant either "Stop what you are doing," "Not now," or "Try something else." When we played finding games, I used "no" the same way I would "cold" or "getting colder"—"yes" meant "hot" or "getting warmer"—in playing a finding game with a child. When Chaser heard "yes" during these games she knew she was looking in the right area or heading in the right direction to find the hidden object. When she heard "no" she knew she had to look somewhere else.

Chaser understood the current "no" correctly as "not now." That only spurred her determination to inveigle me into playing with her. When I looked over at her again, the pile of toys between us was twice as big. She had found seven or eight more of them in the living room, the bedroom, and the loft area upstairs. I recognized Mickey Mouse (there was no mistaking the Walt Disney character doll), Hobby Horse (a stuffed white horse with a blue mane and tail and with legs that ended in a rocking chair–like base), and Nosey, who would have been named Bullwinkle except that Chaser already had a slightly different Bullwinkle the Moose stuffed animal. I couldn't recall the names of the others, but I could have found out if I'd been willing to get out of my chair to go read the names I'd written on them in permanent black marker. Chaser stood behind the toys and looked from them to me while wagging her tail solicitously at medium tempo.

"No," I said. "Be patient, sugar."

"As soon as this program is over, we're going to have to play the cleanup game," Sally said with a smile.

We both turned our attention back to the *Nature* episode, which was using Snowflake's almost forty years in the Barcelona Zoo, from 1967 to 2003, to chart the development of scientific knowledge of gorillas. During that period, the Barcelona Zoo provided the young Snowflake with a gorilla companion of the same age and later gave him an opportunity to mate and help raise a gorilla family (he had twenty-one children, none of them albinos, with different mothers). The zoo also strove to upgrade Snowflake's first bare enclosure into a habitat that resembled his and his gorilla companions' natural environment. These were milestones not just for him, but also for the care of gorillas in zoos around the world.

The next time I glanced in Chaser's direction, she was lying behind the toys, nose pointed straight to them and me in classic Border collie fashion. She raised her head hopefully as soon as I looked at her.

"No, Chaser, not now. Later," I said.

Ten minutes after that there was a soft squeak. Sally and I turned our heads and saw that Chaser had Choo Choo in her mouth. She squeaked it again to encourage one of us to play with her.

"No, Chaser," I said more firmly. "Time out."

Chaser knew "time out" was serious, and she lay back down behind the toys she had assembled. A few minutes later she padded upstairs, and I figured she was looking for more toys to add to the flock on the living room floor. But I kept my eyes on the television. The next thing I knew a blue racquetball bounced down the steps, caromed off the inside of the front door, and wound up in the small trash can next to Sally's desk.

Sally and I both looked up at that. Chaser stood at the top of the stairs looking down at us, her tail wagging rapidly. The documentary still had a few minutes to go, and I let out a sigh of exasperation.

"She can get her own ball, Pill," Sally said. "Chaser, come get Blue."

That was Sally's characteristic reaction whenever Chaser knocked a ball under the furniture, had it bounce into a trash can, or otherwise got a toy snagged somewhere. If she really thought Chaser couldn't reach it, Sally helped her. But otherwise she encouraged Chaser to get it for herself—"You can do it"—and waited her out. I did the same, but I was much less patient, and Chaser counted on that.

"She's just trying to lure you into playing with her, John. Let her get it for herself," Sally always said. We sometimes jokingly called Chaser a con artist, but she never had a hidden motive. Her behavior was always an invitation to play.

My impatience got the better of me. I retrieved the racquetball, beloved by Chaser for its super bounciness and perfect size for catching in her mouth, and went to the foot of the stairs and tossed it up to her.

Walking back to my chair in front of the television, I heard the ball bouncing down the steps again.

"Chaser, what in the heck are you doing?" I said. "Can't you hold on to your ball?"

This time the ball caromed back and forth between the bottom of the stairs and the front door before coming to rest in between them. I looked up at Chaser, who immediately wagged her tail vigorously and craned her neck to look at me. With another sigh I walked back over to the stairs, picked up the ball, and tossed it up to her again.

"Hold on to it this time," I said.

She immediately dropped the ball and it bounced down the steps to me. I caught it on one of its bounces and looked up. Chaser wagged her tail triumphantly — it is amazing how many different shades of meaning a tail wag can convey — and I found myself chuckling.

"This is a new game, is it?" I asked. I tossed her the ball, and she caught it in her mouth. Then she held it in her mouth, grinning on either side of it. I waited a few seconds, but she was teasing me now and I knew it. "Well, if you're not gonna throw it," I said, and turned from the stairs to the living room.

Chaser instantly dropped the ball onto the steps and bowed her forelegs in anticipation of my catching it. I missed the scenes of the documentary on Snowflake's final years. But I was witnessing a wonder of nature in my own house: my dog's invention of a game for us to play. Motivated by boredom and restlessness, which studies show are fertile conditions for creativity in human beings, Chaser had imagined the game entirely on her own, so far as I have ever been able to figure out. I could identify no behavior on my part that prepared her to stumble onto the idea of playing with a racquetball like this. But she knew how bouncy the racquetball was and how she liked to catch it on the bounce, and in the preceding weeks and months she apparently had observed balls rolling down those steps and bouncing off them by accident.

Chaser's brainchild became her favorite indoor game, and she has never tired of it. She enjoys it as much on the steps of Debbie, Jay, and Aidan's house in Brooklyn as she does at home in Spartanburg. Early on she varied the play in two main ways. Instead of dropping the ball onto the steps, she sometimes barely nosed it into a roll over the top step. Either way, the bounce of the ball off the steps was unpredictable, sometimes high and sometimes low. Her bright eyes, pricked-up ears, and wagging tail showed that my bending, turning, and twisting in one direction or another to catch the ball delighted her. She also sometimes zoomed down the stairs as soon as I caught the ball — at first I thought she was finished with the game — so that I could go to the top of the stairs and drop the ball down to her. That added to the fun because it gave her chances to catch the ball on its unpredictable bounces down the steps. She caught the ball, raced up the stairs to give it to me, and raced back down to catch it again several times before coming up more deliberately to change places again. This went on for as long as I cared to indulge her, sometimes with her initiating a change of roles and sometimes with me doing so.

All dogs show remarkable creativity in trying to engage our attention and get us to interact with them. This trait must go all the way back to dogs' first finding a place beside a human fire. This can be problematic for dog owners who aren't prepared for their dogs' drive to interact with people. As child psychologists and expert dog trainers alike know, negative attention is better than no attention, and dogs, like children, will be conditioned to misbehave if that is what most makes people take notice and interact with them, even if the interaction is far from positive. The only lasting cure for that, in the case of a child or a dog, is to react neutrally to misbehavior and strongly reinforce positive behavior.

Misbehavior wasn't on my mind as I marveled at Chaser's invention of a new game. Instead I was thinking of the scenes of Snowflake that Sally and I had just watched. As a highly social animal, Snowflake had a biological need for companionship. Deprived of that in the wild (his albinism would probably have meant a short life there, because his eyes' extreme sensitivity to light affected his coordination and ability to forage for food), Snowflake found a first friend in the primatologist who saved his life. After that the keepers at the Barcelona Zoo seemed always to have treated him with affectionate care, and finally he enjoyed good relationships with other gorillas.

Dogs' social nature is equally strong, but there is a difference. For Snowflake, human companionship could never fully substitute for close interaction with other gorillas. But domestic dogs have evolved to enjoy a unique interspecies social relationship with people. For Chaser, living with people isn't a poor substitute for companionship with other dogs. Being in harmony with a human family is a true fulfillment for her, as it is for pet, working, and service dogs everywhere.

As a Border collie, Chaser's daily requirements for attention and activity are at the high end of the spectrum. Sometimes this can be a little exasperating. Hungry for playful interaction, Chaser will sometimes nose a newspaper right out of Sally's or my hands. Whatever we are doing, she wants to be involved.

But Sally and I have welcomed that behavior for its own sake. And I have eagerly recruited Chaser's immense energy and need for activity with people to propel her language learning. With every passing day, Chaser has demonstrated to me how dogs' social intelligence enables them to learn to do new things with and for people.

By the time Chaser was ten months old, I was no longer able to keep all the objects in her flock straight in my own mind. I frequently needed to look at the name written on a toy if we hadn't played with it in a while. But Chaser had the proper noun names of the more than three hundred objects in her flock, along with mental images of each object, lodged securely in her long-term memory.

In his critique of the Rico study, Paul Bloom granted that Rico could have more than two hundred object names in his long-term memory. But because Rico's knowledge of these object names was tested only by asking him to fetch the objects, Bloom questioned whether Rico understood that an object's name referred to the object apart from this one action.

Bloom noted that children learn that "words are symbols that refer to categories and individuals in the external world." He offered the example of the word "sock," which "even one-year-olds appreciate . . . refers to a category, and thereby can be used to request a sock, or point out a sock, or comment on the absence of one." If Rico treated the sound "sock" only as a one-word signal to "bring-the-sock," then his correctly fetching the sock would "have nothing to do with human word learning."

Bloom was touching here on the fact that children acquiring their first language achieve both referential understanding (knowing what a particular word refers to) and combinatorial understanding (knowing that there are different kinds of words, such as nouns and verbs, that can be put together in many different ways). Although they aren't conscious of it, combinatorial understanding brings children into the realm of grammar, of syntax and semantics. Syntax is the set of rules, which vary language by language, for putting words together in phrases and sentences, such as the command "Fetch the sock." Semantics is about how the meaning of a phrase or a sentence depends on those rules.

Aware of Bloom's critique, Kaminski and her coauthors reported the anecdotal evidence of Rico's owners that he could do more than "fetch the sock": he could also "put an item into a box or . . . bring it to a certain person." But Kaminski et al. acknowledged that they didn't test Rico for these abilities, and Bloom very reasonably said this created doubt about Rico's having referential understanding of any of the object names he knew. Rico might well have learned an association between all the object names and fetching the objects, but no more than that, Bloom argued.

When Chaser was eight months old, I began informally testing her ability to combine object names with more than one command, and to understand independent meanings for names and commands that I paired at random. In effect this meant testing Chaser's capacity for combinatorial understanding of two elements of syntax: object name and action, or noun and verb.

Chaser already knew how to take an object in her mouth, nose it, or paw it on command. I took three newly learned objects — Wise Owl (a stuffed owl with oversize eyes), Punt (a small stuffed football), and Mallard (a stuffed duck) — that I had never asked her to take in her mouth, nose, or paw. I assigned each object and each command a separate number, and then paired them at random.

Wise Owl, Punt, and Mallard were on the living room floor a few inches from each other. With a list of my random pairings of toys and verbs in my hand I said, "Chaser! Take Punt." She immediately picked up Punt and looked at me with her tail wagging, pleased to have the plush little toy in her mouth.

"Good dog!" I said. We played with Punt briefly, and then I replaced it on the floor with Wise Owl and Mallard. I glanced at the next random pair on the list and said, "Chaser! Take Mallard." In a flash she had Mallard in her mouth and was squeaking it happily.

When Mallard was back on the floor with the other toys I said, "Chaser! Nose Punt." Chaser nosed Punt, rolling it across the living room floor, and she ran after it to nose it again, rolling it further, her tail wagging vigorously. This was fun!

Random pair by random pair, I asked Chaser to take, nose, and paw the three toys in different combinations. She made one error in trials of fourteen random combinations, pawing Mallard when she should have pawed Wise Owl. In other informal tests over the next few months, always with objects that I hadn't previously asked her to take, paw, or nose, her error rate quickly dropped from one or two mistakes per test to zero. After that she never made a single error.

This was strong evidence of referential understanding of independent meanings for two elements of syntax. Because I always tested her with newly learned objects that I'd never before asked her to take, paw, or nose, the evidence also showed that Chaser, just like children, could understand novel combinations of two-word phrases. Double-blind testing was necessary to make the evidence incontrovertible, but by the time Chaser was a year old I had no doubt that it would do so.

Bloom's emphasis on children's ability to learn that words can refer to categories of things in the world was important for another reason. Words that refer to categories are common nouns, whereas names are proper nouns that are meant to refer to a particular thing, object, place, or individual. One-to-one associations are sufficient to understand what proper nouns refer to, but common nouns involve what linguists call one-to-many associations, or mapping. And understanding how a proper noun and one or more common nouns can refer to the same thing, place, or individual, like a toy named Uncle Fuzz, involves many-to-one mapping.

Along with many other animals, dogs have some instinctive categories, such as male or female, food or not food, and dog or not dog. Dogs and many other animals can also acquire understanding of categories that are not instinctive, such as categories of visual and auditory stimuli that define the appearance and sound of dogs, other animals, or people. Bloom was skeptical about Rico's referential understanding of a common noun like "sock" because no one had ever shown that dogs (or any other animals) can form categories identified by words. The ability to generalize a single word across an entire category of things, as linguists put it, is one of the most important stages in childhood language learning and one of the most important indicators of higher cognitive function.

I began teaching Chaser common nouns during her first days with Sally and me. When Sally and I trained Chaser to "go to grass" if she heard a car coming, we did this in many different spots around our neighborhood. I also asked her to "go to grass" on our walks when there wasn't a car coming. She soon learned that "grass" could refer to grassy areas anywhere. Likewise she learned that "car" meant a vast array of vehicles of different sizes, including our own, which could be moving or stationary. On walks I also often asked Chaser to "go to trash can." Because of city rules, the trash cans in our neighborhood all looked alike. Or I asked her to "go to tree," based on the shared physical characteristics of trees, despite one tree's often looking very different from another in various ways. "Stick" was another common noun I taught her early on, followed by the ability to distinguish a "big stick" from a "small stick."

The most abstract common noun category Chaser learned in her first three years of life was that of "toy." She began learning the difference between toy and non-toy early in her first year, once she understood the meaning of "no." We always left several toys, including chew toys, out on the floor for her, rotating the toys with those stored in Rubbermaid containers. However, we didn't want Chaser mouthing or chewing normal household objects such as shoes or purses that might be left lying around, or things that tend to wind up on the floor now and then such as towels or wash rags. To handle that I put four objects that belonged to Sally and/or me on the floor in the living room in front of the television with Chaser watching me. If the object was Sally's I held it up and said, "Chaser, this is Nanny's. No play. No play. This is Nanny's." If the object was mine I identified it as "Pop-Pop's."

Each time I identified an object this way, Chaser moved back a little as if to say, "I know that." Sally and I left the objects on the floor in the living room for several months. They got knocked around, or picked up when we were cleaning, but I always put them back on the living room floor in front of the television. Chaser quickly learned what she could and could not play with, and there were only a few instances of Chaser's chewing on a shoe or a belt. Whenever she did, one or two corrections with "No, Chaser" were enough to train her not to chew on it again. After that I could put down new objects that I identified with the "no play" label and she would never touch them at all.

Toward the end of Chaser's first year I built on this early obedience training to teach the difference between toy and non-toy. I took items that looked like many of Chaser's toys but that belonged to Sally or me and had previously been stowed out of reach or out of sight, such as little stuffed animals or other mementos. I held them up and said, "Chaser, this is Nanny's; no toy, no play," or "Chaser, this is Pop-Pop's; no toy, no play."

Again it took only a few corrections, this time with "No, Chaser; no toy, no play," for Chaser to leave these objects alone on the floor. And she could soon reliably fetch a toy from a group of objects that included toys and non-toys with shared physical characteristics. She successfully distinguished between toys and non-toys based on their functional characteristics. Toys were objects she was given to play with, and non-toys were everything else. That meant I did not have to say "no toy" for every object that Sally and I wanted her to leave alone. Chaser knew that only objects I introduced by pointing to them and saying "Chaser, this is ____" were toys. Anything else was a non-toy.

As Sally and I celebrated Chaser's first birthday, I was pumped up about her progress with proper nouns, common nouns, and combining nouns and verbs. With more work in all three areas, I felt sure there would be some very powerful findings to publish.

In the meantime I was excited, and a little nervous, about a demonstration by David Johnson that Chaser and I were going to attend at Wayne West's farm in two weeks. It would be our first visit to Wayne's since the previous November, when the ornery ewe charged Chaser and head-butted her onto the ground. How was Chaser going to react to seeing sheep again?

10

—

Herding Sheep

EIGHT OTHER BORDER collie owners were at Wayne's farm
with their dogs. They were a mixture of full-time farmers and
folks who had other occupations but kept some cattle or sheep
on their property. What the other owners and I had in common was
that we had run into problems getting our dogs comfortable herding
livestock. Some of the dogs were too aggressive and scattered livestock
in all directions. Others were too hesitant and could not get livestock
moving at all.

It was a friendly group of people, and we milled around for a bit
getting acquainted in the warm May sunshine. Wayne's more than two
dozen birdhouses, perched high on poles, were full of purple martins,
members of the swallow family, back from wintering in South America, and their throaty *tchew, tchew, tchew* calls were a pleasant murmur
in the background. Wayne proudly invited us all to come back in late
summer and see how bug-free the purple martins made his backyard
and pool area.

Following her custom, Chaser was delighted to make friends with
the people and had no interest in the other dogs. In that group of
dogs, her behavior wasn't unusual. Although the other dogs showed
some interest in one another, they were all from good working Border collie lines — the majority of them, like Chaser, were from Wayne's
farm — and they were much more interested in the sights, sounds, and

smells of Wayne's horses, cattle, and sheep. It reminded me of when I sat around the campfire with Wayne and other Border collie people and their dogs after the herding trials at the Spartanburg County Fair. The dogs were focused on the people and the exciting work they did together. Other dogs were of little interest in comparison.

Introductions finished, we moved to a large dirt corral where Wayne trained his cutting horses and sometimes put on small rodeos or herding demonstrations for groups of visitors. A stand of bleachers at one end of the corral could seat several dozen people. The other dog owners and I sat down in the first row, and our dogs lay on the ground at our feet, their eyes drawn to the five sheep milling about in the corral.

With David standing beside him, Wayne said, "Y'all know I learned most of what I know about Border collies from Arthur Allen. All things considered, I reckon I learned pretty good. But if there's one fellow who might've learned from Mr. Allen even better than me it's the son of a gun — I mean gentleman — next to me. I wish I knew how he done that, plus never gaining a pound since I met him all these years ago, despite outeating me every time we break bread together. But without further eloquence"— the grin Wayne had been stifling broke out on his face —"I give you Mr. David Johnson. David, you have the floor — I mean corral."

David, as tall as the burly Wayne, but spare and rangy in comparison, blushed and shook his head at that. And then he said, "Wayne, if you could bottle that snake oil, you'd be rich."

That broke us all up, including Wayne, who said, "Matter of fact, I'm working on the formula, and I could use some investors."

As our laughter subsided, David got down to business. He spoke briefly about the need to respect a dog as an individual and wound up, "Learn who your dog is, what he knows and doesn't know, what his temperament is, and then train based on that. Now, I've been watching your dogs carefully since you got here, so let me see if I can show you how to get them started off right with the sheep."

Over the next three hours, David took each of the eight other dogs into the corral with a twenty-foot line attached to a light harness and worked with the dog for fifteen or twenty minutes. As their owners had reported, some of the dogs were too aggressive, others too timid, in trying to move the sheep. With subtle, precise handling of the line and a few words here and there, David firmly but gently restrained the aggressive ones and emboldened the timid ones. After he brought each dog out of the corral, he answered questions about what he'd done.

It was fascinating for all the dog owners to see how David interacted with the various dogs, modeling how best to work with them as individuals. But of course it was especially interesting for each person to see how David worked with his or her dog in particular. David was teaching a master class to two sets of students, dogs and people. I kept wondering how he would work with Chaser and what she and I could learn from that.

Was Chaser ready?

When he'd invited me to the demonstration, Wayne had told me, "Chaser might still be a little young for David to work with, so you should probably plan on her just being an observer alongside of you. She'll learn a lot from that." But the more I watched David with the other dogs, the more I wanted him to work with Chaser, too. Chaser's behavior made me think she was ready: she never took her eyes off the sheep, and each time a dog moved the sheep her ears went up a bit and her tail started wagging. When David brought the sixth dog out of the ring, I managed to ask Wayne and him, "I know you weren't planning on it, but I sure would love it if Chaser could have a chance with the sheep. Is that possible?"

Wayne looked to David, who nodded his head yes. That excited me, but it also made me a little anxious. Was I still rushing? Was Chaser ready for this? It was hard to sit still as David worked with the seventh and eighth of the other dogs.

Finally, it was Chaser's turn. David first took her into the corral without putting the harness and long line on her. For a few minutes he just observed her as she sniffed the ground with no apparent interest in the sheep. But as he later told me, David saw that she was interested, albeit uncertain about what to do, from the way she angled her body slightly toward the sheep rather than away from them. He called her by name, put the harness and long line on her, and walked with her toward the sheep. For several minutes David simply stood holding the line slackly in one hand while Chaser sniffed the ground, but with her eye always on the sheep. I was seesawing between hope and disappointment, afraid that David was going to bring Chaser out of the corral and tell me she wasn't ready.

Then I heard David whispering softly to Chaser, "You can do it. You can do it. You can do it."

I couldn't detect any reaction on Chaser's part. Again I feared that David was going to give up on her. But he kept whispering, "You can do it. You can do it. You can do it."

I couldn't tell you how long he did that, whether it was three minutes or ten minutes. It felt like an eternity.

And then Chaser shifted her feet and pointed her nose straight at the sheep. Her ears went up and her tail began to stand out behind her. David was still whispering, "You can do it. You can do it. You can do it."

The sheep noticed the change in Chaser's position and body language and they began shifting their own feet. At that, Chaser stepped briskly toward them and they began to move off away from her.

David sang out, "Good girl, Chaser. That'll do," and she trotted over to him. Maybe I was projecting my desires, but she looked as pleased as when she'd found a toy I'd hidden. She seemed to be bursting with pride, just like I was. My other feelings were relief that Chaser had made this breakthrough, and admiration for David's intense communion with her.

David brought Chaser out of the corral, and she basked in my pets and praise. David told me, "She'll train up fine, if you've a mind to do it."

What a day it had been. I couldn't wait to tell Sally the good news.

Six months later on a crisp October afternoon I said, "Chaser! Let's go to the sheep."

In a flash Chaser was at the door, tail wagging, waiting for me to come open it.

Sally was going through the mail at her desk along the long wall of the living room. She looked up and said, "You said the magic word, hon. Chaser's raring to go."

I pulled on my jacket and circled through the living room so that

Sally and I could kiss each other goodbye. A few minutes later Chaser and I were in the car heading for Wayne West's farm, about half an hour away. We were in the regular family car, not our pickup, so Chaser was lying on the back seat. But as always when we were driving somewhere, she was having a snooze, resting up for whenever there would be something interesting to do.

Not that she was in any doubt about where we were heading. Though Rico's critics said dogs do not learn words by overhearing them, there were quite a few words that Chaser seemed to have learned in exactly that way. I never tried to teach her the word "sheep," for example. She learned it simply by overhearing it in connection with going to Wayne's.

After David Johnson guided Chaser through her first successful encounter with sheep, Wayne told me, "Carry Chaser on out here with you whenever you got time, Doc. Don't bother calling ahead. If I'm not around, you know where the sheep are. You and Chaser'll do me a favor. Those darned sheep know my dogs so good, and know to heed them so fast, that people don't always appreciate my dogs when I'm demonstrating with them. If Chaser moves those sheep around, that'll freshen up their reactions for the next time I demonstrate with my dogs."

Since then, we'd been going out to Wayne's once or twice a week. It wasn't often enough for Chaser to become really accomplished at sheepherding. For one thing, Border collies can't excel without regularly facing fresh challenges with unfamiliar sheep. Like human athletes competing against more challenging opponents and human students tackling more difficult subject matter, Border collies need to make their way through progressively more difficult situations and problem solving to achieve mastery as working dogs and become champions in herding trials. The same is true of shepherds. People such as Wayne West and David Johnson can guide and partner with dogs so well because of all their experiences with different sheep in varied places and conditions.

But even if Chaser wasn't going to be ready for sheepherding trials anytime soon, we both loved working the sheep once or twice a week for thirty or forty minutes at a time. Over the past six months we'd improved a lot, individually and as a team, in reading the sheep and moving them from one place to another in the good-size pasture Wayne had close to his house. It was fun to watch Chaser growing more and more comfortable at keeping her eye on the sheep while giving me her ear and the occasional glance. The hesitancy she initially showed around the sheep was long gone, and any resistance from the sheep was a welcome challenge. It seemed to me that her tail never stopped wagging as she darted back and forth around the stock in response to my signals.

The visits to Wayne's were rewards to both Chaser and me for her language learning progress, but they were also part of that language learning. I always wanted to keep Chaser's understanding of words grounded in the herding behaviors that are her genetic birthright. Chaser was eighteen months old and so far had learned seven hundred proper noun object names. She easily learned one or two new proper noun toy names a day without forgetting any of the old ones. So I had no doubt that she was going to reach the thousand mark.

About a mile from Flint Hill Farm, Chaser sat up expectantly. It probably took a few drives out to Wayne's for her brain to map the route exactly. But it seemed that she almost instantly knew every straightaway, turn, bend, undulation, and bump of the drive, and tracked it subliminally while she dozed. Yasha and our other dogs showed the same anticipatory behavior in the car as we neared favored hiking and white-water spots, and most dog owners who regularly take their dogs on certain outings have probably experienced this phenomenon. Dogs' apparent ability to map a driving route in their minds may not be a match for the marvels of homing pigeons and migratory birds, but it's impressive in its own right. Chaser always woke up and began to look out the window for Wayne's farm at the same spot along the road.

"You're right, Chaser. We're almost to the sheep," I said. At the word "sheep" Chaser barked happily.

Nobody was at home but the animals when we pulled into Wayne's driveway and parked next to his garage. The dogs in the kennel greeted the car with a cacophony of barks. The barking grew frenzied when I opened the rear door of the car and Chaser jumped out and sped over to the gate to the sheep pasture on the other side of Wayne's swimming pool. All of the dogs wanted to be in that pasture working the sheep, but the most frenzied barker was a female dog, mostly black with white patches, that I didn't recall seeing before.

As soon I opened the gate, Chaser shot through it and ran down the sloping pasture to find the sheep. She disappeared over a little knoll as I closed the gate behind us. I walked to the crest of the knoll and saw that Chaser had Wayne's small flock of five sheep in the far corner of the pasture. She was running back and forth, trying to move them out of the corner.

"There," I sang out, and Chaser stood still.

"Way to me," I said. To follow that signal and go behind the sheep in a counterclockwise direction, Chaser had to move along the fence that formed the right side of the corner the sheep were in. As she did so, the sheep spilled out of the corner.

"Come by," I said, to turn Chaser back behind them clockwise.

"There," I said, when Chaser was directly opposite me and about ten feet behind the sheep. She stopped instantly. "Drop," I said. She went to her belly. I watched the sheep for a second or two as they clustered a little closer together.

"Walk up," I said. Chaser sprang to her feet and began to walk toward the sheep. The sheep moved up the slope to me, and I turned and walked a good seventy yards away to another corner of the pasture, where the road to Wayne's house intersected with a side road.

I turned around to see the little flock of five sheep, all full-grown adults much bigger than a Border collie, coming toward me. Right in among them was Chaser, as if she were part of the flock. Wayne told me he'd never seen another Border collie do that.

One of the sheep broke ranks and tried to peel back the other way. Without a word from me, Chaser darted out from among the other sheep and turned the stray back into the flock. Then she inserted herself back in the group and trotted along again in the middle of them, ears up, tail wagging, and tongue lolling out of her mouth.

All was well in our world. In her own distinctive way Chaser was doing what she was born to do. And I was a proud and happy daddy watching her.

When the sheep were ten or fifteen feet away from me I said,

"Chaser, that'll do." She trotted forward out of the little flock and came to my side, and I praised her for her good work. Without Chaser to push and accompany them in my direction, the sheep stopped and milled around where they were. It was a little after two o'clock now, and I thought we'd move the sheep in another direction for a bit before heading home.

A black and white shape streaked by me on the right. It was the unfamiliar dog from the kennel. Somehow she had gotten loose and then jumped the pasture fence.

I called out, "There," the command to stop, but the dog ignored me. She got behind the sheep and began rushing them toward Chaser and me in the corner. "There," I called out again in a firm voice, but the loose dog kept moving the sheep toward us in a corner. She too was doing what she was born to do, move the sheep to the farmer, even if she wasn't going to give her ear to that farmer.

The kennel escapee nipped one of the sheep in the heel, and they all rushed away from her a little faster. That would have been fine with me, except for Chaser's being by my side. If the sheep crowded in on us, I would be okay. But Chaser might get knocked down and trampled. She could wind up with a broken leg or worse, much worse.

The sheep were almost on top of us. There was no point in giving the other dog more commands she wouldn't heed. I took off along the fence line for the gate, running as fast as I could. Chaser stayed right with me, although she could have raced ahead or peeled off into the center of the field.

The mostly black dog responded by pushing the sheep even harder behind us. It was a miniature stampede. But its consequences could be full-size.

I kept running, and Chaser kept matching my pace. There was the gate, thirty feet ahead.

I glanced back. The sheep were gaining on us.

I sprinted harder, my lungs beginning to burn. Ten feet from the gate I slipped and almost went down. But I managed to hold my balance and keep going.

The sheep were only a few feet behind us now.

Finally here was the gate. I flung up the latch, opened the gate a few inches, and squeezed through it with Chaser. I shut the gate just as the sheep galloped past.

Perplexed, the mostly black dog stopped running at the gate. She clearly couldn't understand why I wasn't holding the gate open for her to drive the sheep through.

I leaned on the gate and caught my breath.

Chaser was fine. She was ready to work the sheep again, in fact. But I figured our close call was enough for one day.

For her part, the other dog was now lying on her belly on the other side of the little flock, holding them by the gate as she assumed the shepherd — me — wanted. I opened the gate and went into the pen, leaving Chaser outside it. The other dog still wasn't willing to heed my commands, but she let me come up to her, and I took off my belt to use as a leash. With that she walked dutifully beside me out of the pen and back to the kennel.

Up close I could see that she was a young dog, perhaps around Chaser's age. Although the pattern wasn't identical, the kennel escapee's mostly black coat with white patches struck me as a reverse image of Chaser's mostly white coat with black patches. I thought they would make a great picture if they were posed side by side.

There was nothing wrong that I could see with the fence around the kennel yard. The dog must have jumped it, driven by the frustration of seeing Chaser go into the sheep pasture to do what she was longing to do. When I put her back inside the kennel yard, she seemed willing to stay there. I was about to put her into the kennel proper, where she couldn't possibly get out, when Wayne drove up in his truck.

Hearing about the dog's behavior, Wayne shook his head and then apologized. With a wry smile he added, "I reckon we could chalk this up to sibling rivalry."

To my astonishment he explained that the escapee, Kate, was Chaser's littermate. One of Wayne's daughters, Sandy, had taken Kate as an eight-week-old puppy, only to find that she was a nipper. With a young baby in her family, Sandy wasn't able to work with Kate. Wayne had worked with Kate himself, and he had recently sent her to David Johnson for additional training. Kate promised to be a good herding dog, and Wayne was looking for the right situation for her.

Sally was upset when I told her what had almost happened, concerned for me as well as Chaser. As we talked over the incident she realized that I had never been in any real danger. And we both laughed about Kate's being Chaser's sister, recalling some of the more raucous moments between Robin and Debbie when they were young.

The question was what to do about Chaser. Should I stop taking her to Wayne's? By dinnertime Sally agreed that was too extreme a reaction. Kate's jumping the kennel fence was just one of those things that can happen now and then in the best of circumstances. But Sally made me promise to make sure all of Wayne's dogs were secure in their kennel and keep an extra sharp eye out for any escapees. That reminded us both of those great escape artists Blue and Timber, and we shared a few stories of their misadventures over dinner.

Three days later Chaser and I went back out to Wayne's. Kate stayed in the kennel, and Wayne soon found her a good home on a farm with horses and cattle. Since then Chaser and I have continued our regular visits to work Wayne's sheep without any mishaps. Thanks to Wayne, I have been able to keep my promise to Chaser when she was an eight-week-old puppy, that I would strive to help her fulfill herself as a Border collie. I love that she can herd sheep the way she was bred and born to do while continuing her progress in herding words.

In Chaser's second and third years, that meant a greater focus on common nouns, words that stand for categories, and matching another benchmark from the Rico study, learning by exclusion.

11

—

Advanced Lessons

IN CHASER'S SECOND year I put more effort into teaching common nouns. My goal was to teach her two more common noun categories in addition to "toy": "ball" and "Frisbee."

To teach Chaser what a ball is I started with eight balls of different sizes, colors, and materials on the floor: a tennis ball, a racquetball, a baseball, a lacrosse ball, a golf ball, and big and small balls made of foam and spongelike stuff. The balls had many different characteristics, but they were all round and, to different degrees, bouncy. To enable Chaser to start off with errorless learning, there were no other objects on the floor. So when I said, "Chaser, find a ball," there was no way she could choose an incorrect object. I used the 8-of-8 test procedure I described earlier, in which I did not replace objects in the group after she picked them correctly. Instead I kept saying, "Chaser, find a ball," until there were no balls left on the floor.

We did this several times off and on through the day. And then we went through the same procedure with another eight balls of different types, followed by another eight balls, and so on. If this process was successful, Chaser would learn to generalize that balls are round and bouncy. She would acquire an abstract concept of what a ball is, based on the common physical characteristics of all the balls she encountered.

After many repetitions of this over several weeks, I made the task more challenging by putting out eight balls and eight non-balls. She not only had to generalize what characterized a ball, but also had to discriminate a ball from a non-ball. I randomly asked her to retrieve balls and non-balls. If she retrieved a non-ball when she was asked to retrieve a ball, I softly told her, "No, Chaser. That is not a ball." Informal tests soon demonstrated Chaser's ability to bring a ball, and only a ball, when I asked her to do so.

I used the same procedure for teaching her what a Frisbee is, using that brand name to apply to any throwable, catchable spinning disk or ring in her flock of objects, no matter what it was made of. First I put eight Frisbees on the floor, with no other objects in sight. After errorless retrieval of all eight, I tested her on other sets of eight Frisbees so that gradually she generalized the characteristics of a Frisbee across all the throwable, catchable disks in her flock. And then I put eight Frisbees on the floor with eight non-Frisbees, so that she would have to discriminate successfully among the Frisbees and non-Frisbees.

As I mentioned earlier, Chaser learned the concept of "toy" not on physical characteristics, but based on an abstract functional characteristic. Toys were objects she knew by their individual names and could play with. Everything else was a non-toy. It was fascinating that she grasped the higher-level, more abstract concept of "toy" first.

As Chaser's learning continued through the fall of 2006, I was of two minds. All the experiments Chaser and I had conducted so far — proper noun learning, independent meanings for two elements of syntax, and common noun learning — showed that her language abilities reached far beyond those documented for Rico. The evidence was piling up for a major scientific paper in a peer-reviewed journal.

Part of me was itching to document Chaser's achievement for its own sake, as a contribution to the understanding of learning, and as a stimulus to other scholars. The other part of me didn't want to slow down research in order to write such a paper, which would limit the time I could spend with Chaser day in and day out. Instead, I wanted to press ahead and extend Chaser's language training. Her ability to take an object in her mouth, or nose or paw it, on command showed that she could respond correctly to sentences with two elements of grammar: a verb and a direct object. In the early 1980s Louis Herman

Chaser next to some of her toys — she knows the unique names of over 1,000 objects. SEBASTIEN MICKE, PARIS MATCH

Rico gained international attention in 2004 after his owner, Susanne Baus, reported that he recognized the names of more than two hundred simple words. He was studied by the animal psychologist Juliane Kaminski at the Max Planck Institute for Evolutionary Anthropology in Leipzig, Germany. SUSANNE BAUS AND WITOLD KRZESLOWSKI

Clever Hans amazed crowds in early twentieth-century Germany with his apparent mathematical abilities, later shown to be the result of unconscious cuing by his owner, Wilhelm von Osten, shown here. Any researcher who wants to demonstrate animal learning must avoid the Clever Hans effect. COURTESY OF THE NEW YORK PUBLIC LIBRARY

Yasha, my devoted classroom assistant and white-water rafting partner.
COURTESY OF THE AUTHOR

Grindle was a big-hearted German shepherd with a gift for turning doorknobs. COURTESY OF THE AUTHOR

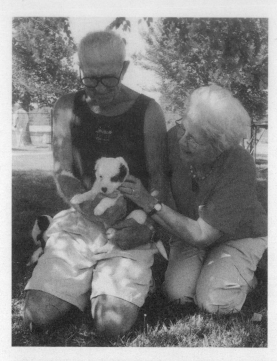

Sally and I meet Chaser for the first time at Wayne West's Flint Hill Farm in 2004. SANDY HANCE

As a puppy, Chaser demonstrated boundless energy and an aptitude for learning. ROBIN PILLEY

Chaser herds sheep as I give herding commands. Often working Border collies know each sheep in a flock by name. ROBIN PILLEY

Some of Chaser's friends at Wayne West's farm. These sheep are frequently used to help train young Border collies. ROBIN PILLEY

Wayne West raises sheep and Border collies at Flint Hill Farm and taught Chaser and me how to herd sheep.

COURTESY OF THE AUTHOR

In testing Chaser's syntax and semantics understanding, I instruct Chaser to retrieve one of two possible direct objects from the bed and take it to one of two possible prepositional objects in the living room.

COURTESY OF THE AUTHOR

Each of Chaser's 1,022 toys has a unique proper noun name.
ROBIN PILLEY

Chaser has demonstrated the ability to deduce the name of a novel object when it is placed among a group of familiar toys.
ROBIN PILLEY

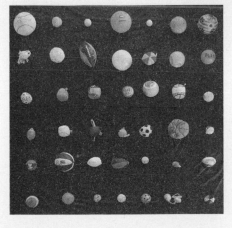

Chaser knows the proper noun names of over one hundred individual balls, which she can also identify by the category labels "ball" and "toy." ROBIN PILLEY

My former student Alliston Reid (on the far left) collaborated with me on the paper we published in *Behavioural Processes*.

Play is a very effective reward for Chaser. One of her favorite games is playing fetch on the stairs . . . and most of the time she makes me do the fetching.

and his colleagues at the University of Hawaii studied the ability of two bottlenose dolphins, Phoenix and Akeakamai, to respond correctly to sentences containing two, three, four, and five elements of grammar, including verbs, direct objects, indirect objects, prepositions, and adjectives and adverbs. Understanding three elements of grammar sentences was the next step for Chaser in my mind.

Working alone, I had to make a choice. If I didn't write a paper now, who knew what might prevent me from doing so in the future? Ultimately I couldn't take the chance of not documenting Chaser's learning as soon as I reasonably could. It would be a terrible waste if her learning wasn't shared with others.

More important than any concern I might have about being able to write a paper later, however, was my desire to honor Chaser's achievements. By this point she had achieved nearly all of my initial research goals for her, matching or exceeding Rico's learning and doing so in a way that met the objections that the Yale research psychologist Paul Bloom and two other childhood language learning researchers, Stanford's Ellen Markman and Maxim Abelev, had raised about the Rico study. These objections all involved language understanding that one-to-three-year-old toddlers display but that had never been demonstrated or was subject to question in animals.

In assessing the Rico study, Bloom emphasized that children learn words by overhearing them, the implication being that animals never did this. Yet it seemed plain to me that Chaser had learned "sheep" and quite a few other words by overhearing them. All of my experiences with dogs over the years made me suspect that many dogs commonly understood the meanings of a dozen or more words they heard frequently, such as the obedience commands or the words "walk" and "treat."

In addition to "Let's go to the sheep," Chaser responded enthusiastically to "Let's go for a hike," or "a walk" or "play Frisbee." If Sally said, "I'm going over to Sue's" or "I'm going over to Nora's," Chaser headed straight to the door. As soon as Sally opened the door, Chaser ran to the edge of the road and waited there for her.

There were also words she learned by overhearing that she preferred not to hear. If we said, "Chaser, let's go to the store," she looked peeved and didn't want to come. Her ears and tail went down, and her usual grin turned into what looked like a slight frown. She came along reluctantly only if we pressed her. She had learned that going to the store meant sitting in the car waiting for one or both of us to return from shopping. Where was the fun in that?

Another thing Chaser didn't like was taking her monthly heartworm pill. If Sally said, "Chaser, come get your pill," she slunk away to the bedroom and tried to avoid it. Sally got around that by saying, "Chaser, come get your pill and I'll give you a treat." That brought Chaser quickstepping to swallow her pill dutifully and then enjoy a treat.

The real problem with overheard word learning by an animal was demonstrating it empirically. To show that an animal's exposure to a word was only by overhearing it, with no explicit teaching, would require a 24/7 visual and audio record of the animal's life to that point. I put that aspect of Chaser's learning to the side in my mind, unsure whether I should include it in a peer-reviewed paper.

Bloom also said that children are able to learn words by being shown an object and hearing a person name it, again with the implication that animals could not do this. Showing Chaser an object and naming it became my basic method for teaching her words. Once Chaser learned the concept that objects have names, as I described in chapter 9, my tests showed that she learned the names of objects on one trial when she was shown an object and told its name. When I tested retention after ten minutes and twenty-four hours, I found that she needed additional rehearsal to process the learning into long-term memory. But toddlers also need rehearsal to lodge new words into long-term memory.

Another of Bloom's reservations about Rico's learning was that "it is always Rico's owner who is communicating with him. . . . Yet, if Rico really is learning sound-meaning relations . . . it should not matter who the speaker is." That seemed reasonable to me. As Chaser's word learning advanced, I brought other people in as trainers. Sally, Robin, and two Wofford undergraduates who assisted me at different times, Caroline Reid and Katie Grainger, all taught Chaser proper noun object names. I was also planning to do a set of formal blind and double-blind tests of Chaser's learning, in which others beside myself gave her the commands to perform language tasks.

Rico was only tested in fetching objects by name, and Paul Bloom's biggest question was whether a dog could show knowledge of words in relation to more than one action. As I've described, Chaser showed this in learning to take named objects in her mouth, or nose or paw them, on command, including objects she'd never been asked to do any of these things with. This demonstrated both that she understood independent meanings for two elements of syntax, an object name and an action verb, and that she understood novel combinations of objects and commands.

Bloom also questioned whether animals could learn common nouns. I was looking forward to documenting Chaser's understanding of "ball," "Frisbee," and "toy" as common nouns.

Markman and Abelev seconded Bloom's reservations and added two of their own. First they were skeptical that language-trained animals could progress beyond "mutual exclusivity," the assumption that things have only one name. Markman and Abelev observed that in the early stages of language learning five-month-old infants "expect that a novel word will refer to a novel object rather than a familiar one," but by seven months most babies learn that this "mutual exclusivity" does not always hold true and that a single thing can have multiple names. They learn that a set of common nouns such as "cat," "kitty," and "pet" can all refer to the same thing, and that a set of proper noun names and nicknames such as "Elizabeth," "Lizzy," and "Sweetie" can all refer to the same person.

When I read Markman and Abelev's article on its publication in November 2004, Chaser was seven and a half months old. At that point she knew that she was also called baby, Puppy, girl, and sweetie, as well as combinations of those names and nicknames. Simply by overhearing the various names, she knew that Sally was also Mom and Nanny, and that I was Pop-Pop, Dad, John, and Pilley. She knew that food had many names: breakfast, lunch, supper, treat, chew stick, and bone. She knew that the command to jump had three labels ("hoop," "up," and "over") and that the command to come had four labels ("here," "come," "hurry," and "right now").

In her common noun learning, Chaser showed that she could distinguish balls and non-balls, Frisbees and non-Frisbees, toys and non-toys. She could also identify one of her balls both by its proper noun name and as a ball and a toy, or one of her Frisbees both by its proper noun name and as a Frisbee and a toy.

Thus Chaser demonstrated the one-to-many and many-to-one mapping of words and things that both Bloom and the team of Markman and Abelev doubted dogs could achieve.

Markman and Abelev's second big reservation about the Rico study had to do with learning by exclusion, also known as reasoning by exclusion. Rico demonstrated this when he retrieved a novel object from a group of familiar objects on hearing its name for the first time. In the absence of any external or physical cue, he apparently inferred that the novel name referred to the novel object. However, Markman and Abelev questioned whether Rico's success really showed that he was drawing an inference.

Markman and Abelev granted that the procedure for Rico's "reasoning-by-exclusion test trials ... closely parallel[ed] those used to demonstrate word-learning by exclusion in young preschoolers." But they raised three possible objections. First they noted that Rico's tested vocabulary only included proper nouns having a one-to-one correspondence with the objects they named. Based on this prior experience, Rico might assume that words are mutually exclusive and that thus a novel name must refer to a novel object.

The second possibility Markman and Abelev suggested was that Rico had a default assumption that words are mutually exclusive based on other prior learning or an innate propensity. This could also make him always pick a novel object on hearing a novel name.

Chaser's combined proper noun and common noun learning showed that she knew that words are not mutually exclusive. She understood that objects could have more than one common noun category label ("toy" as well as "ball" or "Frisbee") plus a unique proper noun name. And she understood that the same individual could have more than one proper noun name. Her demonstration of many-to-one and one-to-many word-object mapping put these first two possibilities to rest, insofar as her own learning by exclusion was concerned.

The third possibility Markman and Abelev suggested was that Rico had a baseline preference for choosing novel objects. They pointed out that the Rico study said nothing about testing whether Rico had such a baseline preference. Studies of learning by exclusion in preschoolers did test for a baseline preference for novelty, and such a preference, rather than a correct mental inference, might be why Rico picked the novel object after hearing its name for the first time.

In response to this third possibility, the Rico study researchers replied that, before asking Rico to retrieve a novel object, they first asked him to retrieve two familiar objects. This demonstrated that his choices were not owing to a preference for novel objects.

I decided to replicate this procedure. In addition, I decided to measure Chaser's baseline preference for novelty before her first opportunity to learn a word by exclusion.

I did not attempt to train Chaser to learn by exclusion. But I felt confident she could do it, based on Rico's example and, more important, her own ability to grasp a series of abstract concepts. Chaser grasped that objects could have names, and then became able to learn a new object name on one trial. She learned that common nouns refer to categories, and then became able to distinguish categories based on shared physical characteristics (balls versus non-balls and Frisbees versus non-Frisbees) or more abstract shared functional characteristics (toys versus non-toys).

Before testing Chaser's ability to learn a word by exclusion, I tested her baseline preference for novelty. I took eight familiar objects and rehearsed Chaser in retrieving them by name. The next morning, with Chaser waiting in my study upstairs so that she could not see what I was doing in the living room, I put the eight familiar objects and two novel objects behind the couch. Then I called Chaser downstairs and asked her to retrieve each of the eight familiar objects by name. She did this without a hitch.

Over the next week I repeated this procedure seven times, each time with two different novel objects and eight different familiar objects. Because Chaser now knew about nine hundred proper nouns, there was no shortage of familiar objects.

In all eight trials, Chaser only retrieved the familiar objects I asked for by name. This established that she had a zero baseline preference for novelty.

Next I rehearsed Chaser on retrieving seven different familiar objects by name, in order to be sure that she still remembered their names. After that I had Chaser wait upstairs so that she could not see what I was doing, and I put the seven familiar objects behind the couch along with a novel object that I had not used in looking for a baseline preference for novelty. When I called Chaser to come downstairs into the living room and sit in front of the couch, I first followed the Rico researchers' procedure by asking her to retrieve two familiar objects.

And then I said, "Chaser, find Lounge. Find Lounge."

Chaser padded behind the couch while I sat with my back to her. Several seconds passed without her bringing anything to me. I repeated, "Chaser, find Lounge. Find Lounge." I waited ten more seconds, marking the time by the clock above the television, before turning around to see what Chaser was doing, or rather, not doing.

She was just standing over the novel object and the remaining five familiar objects. I turned back around quickly to make sure I didn't unknowingly give her an identifying cue. And then I repeated, "Chaser, find Lounge. Find Lounge." Over the next minute I repeated those words several times, but she still did not return with an object.

I was wondering how much longer I should wait when Chaser slowly came around the couch. In her mouth was a plush miniature chair from a dollhouse set.

"Good dog! Good girl, Chaser! You found Lounge!" thundered out of my mouth in relief and excitement. The five minutes of play with Lounge were extra joyful for me.

The next day I repeated the procedure with seven different familiar objects and a different novel object she'd never seen before. She immediately retrieved the two familiar objects I asked her to get. And then I said, "Chaser, find Chipmunk. Find Chipmunk."

Again there was quite a long pause, more than a minute altogether, during which I repeated "Find Chipmunk" several times. Again she finally came slowly out from behind the couch with the correct object, a little stuffed animal in the form of a chipmunk.

In subsequent trials, there was no delay. And Chaser made the correct inference every time. After hearing its name for the first time, she always found and brought me the novel object she'd never seen before.

I immediately followed these novel choice trials with retention tests. I placed a novel object whose name Chaser had learned by exclusion behind the couch in a group with three novel objects whose names she hadn't learned and four familiar objects. Immediately after the selection choice, the retention test showed that she remembered the object name. When I repeated the retention test after ten minutes and after twenty-four hours, the results were inconsistent, as they also were with human toddlers.

With this learning, Chaser matched Rico's demonstration of selecting an object on the basis of exclusion. She was in shouting distance of learning more than a thousand proper nouns. And she had met all the additional criteria for word learning proposed by Bloom and by Markman and Abelev. It was time to begin writing up the experiments for publication in a peer-reviewed journal. Given Chaser's unprecedented language learning, I assumed it would be no trouble at all getting a paper published in a good journal.

I should have known better.

12

Getting Published

B Y THE TIME Chaser turned three in the spring of 2007, she knew more than a thousand objects by their proper noun names. There were 800 stuffed animals, 116 balls, 26 Frisbees, and 100-plus plastic and rubber items. I began writing a paper that would share the impressive results of her learning. I hadn't written a peer-reviewed paper in a long time, but I knew the form well. And I thought I had an excellent recent model in the Rico paper, which was distinctive not only for its content but also for its fairly conversational style, though I intended to provide a more thorough explanation of my training and testing procedures.

I included the full spectrum of language learning I observed in Chaser, even if I didn't have extensive data on some aspects of it. I believed that the remarkable nature of the findings would justify publication in journal editors' eyes. I entitled the paper "Can a Dog Learn Nouns, Verbs, Adverbs, and Prepositions?"

My hope was that *Science*, the world's most prestigious scientific journal along with Britain's *Nature*, would publish the paper as a sequel of sorts to the Kaminski paper on Rico. And I hoped they'd ask Paul Bloom to contribute another "Perspective," in which he would acknowledge that Chaser's learning met all the basic criteria for word learning. Toward the end of the summer I told Alliston Reid that I was going to send the paper to *Science*.

With a smile Alliston said, "They only give you one shot there, John."

"I know," I said. "But they published the Rico paper and I've gotta think this is a pretty good shot, given how much Chaser has learned and how she's met Bloom's and Markman and Abelev's major criteria for referential understanding. Here's hoping, anyhow."

A few days after that, with my eyes blurry from reading and re-reading for typos and grammatical errors, I sent the paper to *Science*. Several weeks later an editor at *Science* briefly e-mailed me to say that upon review they were rejecting the paper. There was little detail as to what might have been said about the paper during the review process or what its specific flaws might be.

I was stunned.

Rereading my paper, I confirmed that my experimental procedures in testing Chaser's understanding of proper noun words and her ability to learn by exclusion were identical to those in the Rico study. In addition I had sharpened the paradigm for testing Chaser's exclusion learning by establishing that she had no baseline preference for novel objects. Likewise, I had described rigorous procedures for teaching and testing the learning of common nouns.

But I had to admit that some critical details of my studies were missing. Unfortunately I had not presented the usual tables and figures to display my findings but had only described them in words and a few key numbers. I also recognized that much of the paper was too informal and did not say enough about how Chaser attained her language learning and how I tested it. So I optimistically set about rewriting the paper.

Although I felt my paper was comparable to that of Kaminski and her colleagues, I recognized that as a retired professor from a small liberal arts college and without a long list of peer-reviewed papers to my credit, I had nothing like their stature in the animal science community. At the time of the Rico study's publication Kaminski led a research group at Germany's Max Planck Institute for Evolutionary Anthropology (she has since moved to the University of Portsmouth in England), the equal of any research institute in the world in that field.

In evaluating scientific experiments there are two major errors:

Type 1 errors and Type 2 errors. Type 1 errors occur when you read too much into experimental data. That means you see an effect that has not actually occurred. Type 2 errors occur when you read too little into experimental data. That means you miss a significant effect that has actually occurred. The event is accurately represented in the data, but you don't see it.

As an experimenter, I was more concerned about making a Type 2 error and failing to report Chaser's genuine achievements than I was about making a Type 1 error and overstating those achievements. But scientific journal editors and peer reviewers have to be more concerned about making Type 1 errors and publishing something that won't stand up to scrutiny by the scientific community at large. Those two perspectives constitute one of the constant push-pull dynamics in the course of scientific progress. And science needs both. It needs experimenters to try bold things, and it needs their peers to hold them to a tough standard.

Before long I was bogged down in my data. At seventy-nine years of age I no longer had the head for figures that I did as a younger scientist. Hearing me express my frustration about this one day, Robin said, "You should get Alliston to collaborate with you on the paper. You know he's an expert on all that data stuff."

I would have loved to have his help. Alliston and I always had a great time together, whether we were conducting an experiment in the lab or going on a kayaking outing, and our families were close. Alliston and Robin became good friends when they were both Wofford psychology majors. I officiated at Alliston's wedding to his wife, Leonor, who became one of Sally's best friends, and Sally and I were godparents to their two daughters, Caroline and Rebecca. But Alliston had much more on his plate now than when he was my teaching assistant.

Since graduating from Wofford, he had gone beyond me as an animal scientist in many ways. When he presented his final PhD dissertation data, John Staddon, his thesis advisor at Duke, shook his hand and said, "Welcome to the top twenty." Staddon meant the top twenty in the quantitative analysis of animal behavior. Staddon himself was at least in the top three in the world in that field, which he'd done as much as anyone to establish.

After teaching for several years at the National Autonomous University of Mexico in Mexico City and Eastern Oregon University, Alliston came back to Wofford to fill the spot on the psychology faculty that opened up when I retired. He had a knack for engaging his students in novel ways, such as his use of rat basketball to teach the principles of conditioning, reinforcement, and learning. He guided students in training rats to take a small ball in their forepaws, stand up on their hind legs, and drop it through a miniature basketball hoop and net. Teams of students and their rats competed in Alliston's rat basketball tournament every year.

Alliston was now very much a leader in the quantitative analysis of animal behavior. He was part of a changing of the guard as the researchers of Staddon's generation retired. In fact, Alliston had just started a multiyear commitment as program chair of the Society for the Quantitative Analysis of Behavior (SQAB). Among his responsibilities was editing some thirty papers from scholars around the world for a special issue of the British journal *Behavioural Processes*. That work, his teaching load at Wofford, and his own research weren't going to leave him much time to collaborate with me, and I didn't think it was fair to ask him.

"I'll think about it," I told Robin.

Fortunately, Robin took things into her own hands. She arranged for the three of us to have lunch together early in the fall of 2007. We hadn't been sitting down long when Robin said in her deliberate way, "I think Chaser's learning is amazing. And, Dad and Alliston, I think the two of you could do a paper together that would have a huge impact. Dad didn't want to ask you because he knows how busy you are, Alliston. But I believe the two of you should partner up and do that."

Robin tilted her face down to her plate and raised only her eyes to look first at me and then at Alliston. Following her eyes to Alliston, I was delighted to see that he was smiling happily and to hear him enthusiastically say, "I'm in."

"Thanks, Alliston. I'm in too," I said.

Unfortunately, it wasn't until the summer of 2008 that we were ready to submit the paper, now titled "Collie Demonstrates Referential Understanding," to the journal *Animal Cognition*. In the months since *Science* rejected the first draft of the paper, Alliston and I had improved it a great deal. I'd rewritten most of the paper and added new passages with details on Chaser's training and testing. As I was doing that I often bounced ideas off Alliston, and he gave me many valuable suggestions for demonstrating that the tests were rigorous. Alliston also guided me in the statistical analysis of my experimental data. Finally, along with the paper itself we submitted a video of some of Chaser's trials.

On November 18, an editor at *Animal Cognition* e-mailed us that the journal was rejecting the paper based on the anonymous peer reviewers' comments, which were attached to the e-mail.

My heart stopped.

And then I saw that the two peer reviewers expressed very different attitudes to the paper.

One peer reviewer concluded, "I cannot recommend anything about this paper. The authors appear to lack an understanding of experimental controls and design in the animal language arena." The reviewer also asserted that we should use "label" to refer to the proper noun names of Chaser's toys rather than calling each such proper noun a "word."

That comment told us that the reviewer followed the school of linguistics researchers who claim that only humans can learn words and that if an animal can learn a term for something, it can't be a "word." I believe this reviewer was looking at the paper through such biased eyes that he or she was unable to appreciate the details of the procedures used to test Chaser's learning.

The other peer reviewer also saw problems with our study, and said it included too many assertions about too many kinds of language learning. Except on proper noun learning, the reviewer said, "there is hardly any systematic data."

However, this peer reviewer added, "[On proper noun learning] the authors provide data which is quite convincing. . . . The authors show that Chase [*sic*] . . . can distinguish hundreds of objects by their label. That is fascinating and a great finding in itself. However, to be able to state that Chase can also use adverbs, categories, verbs, etc., it would need exactly the same kind of systematic investigation, which I hope the authors will provide in a revised version of the MS."

That was more like it. This peer reviewer went on to say that we needed to do a better job of demonstrating that Chaser could learn a word in one trial and learn a word by exclusion, and that throughout the paper we needed to provide fuller details on Chaser's training as well as testing. Finally, this reviewer said we had to clarify our statistical data. All this information was necessary in order to provide other researchers with the information needed to replicate our study.

That the journal editor had given more weight to the totally negative review than to this very encouraging one angered me. It is difficult for a research finding to win scientific acceptance when it deviates from the ruling paradigm in a field. The seventeenth-century mathematician and philosopher René Descartes established one of the strongest paradigms in biology and psychology with his idea that nonhuman animals cannot reason or feel but are more or less machines made out of meat. Two hundred years after Descartes, Charles Darwin stated his belief that animals, and dogs in particular, can feel and express emotions and can reason. But anyone trying to demonstrate language learning by animals still has to battle those who faithfully follow Descartes's ruling that animals are just meat machines.

For a while I focused more on the negative judgments than the positive ones. I felt that we were facing a huge uphill battle to get the study published. Sally and the rest of the family grew concerned as the weeks went by and I still wasn't back to being my usual positive self. Not even the Christmas holidays, which were always a joyful time in our family, managed to lift my mood. My play and training time with Chaser decreased. My television time increased. My social life with my Wofford peers went to zero. I skipped my morning workouts.

My family's efforts finally buoyed my spirits. Sally stepped up her always frequent hugs, kisses, and smiles. Debbie pumped me up with daily phone calls. Robin kept saying that Chaser's learning was a huge deal and that Alliston and I would eventually get the paper published. And Chaser would not rest until Pop-Pop was back to normal and ready to play. Seeing that dropping toys at my feet was not enough to stop me from staring glumly at the television, Chaser put her front paws on my lap and her nose right in my face until I got up out of my chair.

On January 25, 2009 — Debbie's birthday — I got out of bed and decided that this was the day I was going to pull myself up by the bootstraps and get back to work on the paper. The rejections had knocked the wind out of my sails, but now I was ready to accept the rejections and improve the study accordingly. If details were what the peer reviewers wanted, details they would get.

I had to do additional tests of Chaser's learning in blind and double-blind conditions, and record them on video. I also needed to describe my training and testing procedures more fully. Alliston had to display my data in tables and figures, along with statistical analyses of the different studies I had done with Chaser.

Knowing that Chaser's learning was real, I went back to the paper determined to demonstrate it with findings that gave the scientific community a high degree of confidence in her learning. The first tough decision in revising the paper was to limit its scope. I had hindered the paper by putting in my anecdotal observations of all of Chaser's language learning, including adjectives, adverbs, prepositions, and overheard words. Alliston had earlier suggested eliminating this material, but I had wanted to be as comprehensive as possible.

From now on the paper would deal only with the four major experiments where we could offer ample data and demonstrate Chaser's learning in blind and double-blind conditions. The four experiments aimed to demonstrate 1) that in her first three years Chaser learned, and that she still retained, understanding of the proper noun names of 1,022 objects; 2) that she understood the separate meanings of proper noun names and commands and could correctly interpret random combinations of these two elements of syntax; 3) that she understood the three common nouns "ball," "Frisbee," and "toy"; and 4) that she could learn a word through inferential reasoning by exclusion. Together these four experiments matched and extended the Rico study, and satisfied the major word learning criteria proposed by Yale's Paul Bloom and Stanford's Ellen Markman and Maxim Abelev.

Discussing the need for a video clearly demonstrating Chaser's learning of more than a thousand proper noun object names, I said to Alliston, "How about we ask the Psychology Kingdom to host us at one of their meetings this semester, so we can record a demonstration with a bunch of students?"

Alliston said, "I'll ask Dave Pittman."

Dave Pittman was another student of mine who had become a psychology professor at Wofford. He was the faculty advisor for the Psychology Club, which most Wofford psychology majors participated in regularly. Alliston and I—and Dave, too, in our conversations—referred to it as the Psychology Kingdom because of a running joke that began when they were students and I was the club's faculty advisor.

On an outdoor trip with the Wofford Adventure Club in the early 1970s, I was jawboning around the campfire late one night with two students, Eddie Coffey and Chris Harris, who were both psychology majors. Eddie and Chris started riffing on the idea that we should rename the Psychology Club the Psychology Kingdom so people could act out their favorite personas, like being a knight errant or a wizard. I joked, "Well, if there's gonna be a psychology kingdom, I'm the king."

Eddie cracked a smile and said, "Don't agree with him, Chris, or we'll never hear the end of it." But then he added, "Well, Sally has thrown so many great parties for the Psychology Club that she has to be the queen of our kingdom"—he paused a beat and looked at me—"so I guess that makes you the king by default, Doc."

That broke us all up. By the time we saw the last embers of the campfire out, Eddie and Chris decided that they were, respectively, the prince and the grand duke of the Psychology Kingdom. When we got back to campus, they began spreading the word, and we wound up with about twenty self-styled members of the kingdom's court. With a nod to James Fenimore Cooper's novels and thanks to his impressive outdoor skills (studious though he was and looked, he would have been a cinch to win the Hunger Games), Alliston became the Deerslayer. Lest we take ourselves too seriously, however, whenever I was introduced as the king I always replied that I was "King of Fools."

Dave loved the idea of having Chaser demonstrate her learning for the current Psychology Kingdom. A few weeks later, Alliston, Chaser, and I assembled with Dave, a few other members of the psychology faculty, and about a hundred students in an amphitheater-style auditorium at Wofford. We also had all 1,022 objects in Chaser's flock of surrogate sheep in sixteen large Rubbermaid containers.

Dave had arranged for Wofford's official photographer, Mark Olencki, and a local videographer to document the event. While Mark was taking his first photos of the group, Robin suggested that the students should throw Chaser's toys in the air so he could get a dramatic action shot. Chaser enjoyed the spectacle, and she was in her element meeting the students and getting attention from them.

A chaotic scene became even more chaotic as ten students in teams of two rummaged through the 1,022 toys piled on the auditorium floor. The five teams each picked ten toys. The selection took a few minutes because the students weren't content to grab just any toys. They competed to find the most interesting and cutest toys.

The selection of objects was random. In addition, none of the students had ever worked with Chaser and none of them knew the names of the objects prior to selecting them. The teams made lists of the names written on their ten objects and gave the lists to me. Each team in turn then randomly placed its chosen objects on the floor behind me so that I could not see the objects when I asked Chaser to retrieve them by name. These procedures constituted a double-blind test of Chaser's learning of the toys' proper noun names.

The fifty objects in total included some whose names Chaser had learned as a young puppy, some she'd recently learned, and some she'd learned at different points in between. In at least seven cases I was not sure what object a name referred to, because I could never keep all of the 1,022 objects and their names straight in my own head.

Chaser's memory for them was better than that. Following her retrieval of the objects, the five teams of students evaluated her accuracy. She had correctly retrieved forty-six of the fifty objects, or 92 percent. The entire Psychology Kingdom cheered Chaser's success.

Dave Pittman opened up the meeting for questions, and John Lefebvre, newly appointed chairman of the Psychology Department, asked if Chaser would retrieve objects for people other than me. I invited John to come down onto the stage and find out for himself.

John is not bashful, and in a minute he was on stage. Having sat in on several of John's classes, I knew he would test Chaser's limits. I threw eight of Chaser's toys on the floor, and asked John to have at it. He picked the smallest stuffed animal, a dog named Tiny. And then, out of Chaser's sight, he threw Tiny several steps up in the amphitheater. He turned to Chaser and said, "Chaser, find Tiny."

Chaser quickly nosed through the objects on the floor. No Tiny. Although I thought the test was a little unfair because Chaser was in a strange environment and did not know John, I said nothing.

John repeated, "Chaser, find Tiny," and Chaser began to explore the room. Several times she reexamined the objects on the floor. John wisely continued to repeat, "Chaser, find Tiny." At least two minutes passed while Chaser looked all over the stage. Finally she approached the steps that led up through the rows of seats in the amphitheater, saw Tiny, and picked him up in her mouth.

Realizing that I was holding my breath, I heaved a huge sigh of relief. Again John showed his knowledge of dogs by enthusiastically praising Chaser. And the Psychology Kingdom of students clapped and roared their praise as if they were at a Wofford football game.

Dave Pittman was looking at the clock, but before he could bring the meeting to a close, a student asked for one more demonstration of Chaser's word learning. I said, "Chaser has learned the meaning of 'clean up,' so I'm going to ask her to do that now. You will see, however, that she is like a two-year-old child and will need reminding to finish the task." I turned to Chaser and said, "Chaser, clean up." She quickly put two toys in the plastic tub but then dropped one at my feet. "No, Chaser," I said. "It's time to clean up." With that reminder, and a few more before we were done, Chaser put all eight toys in the tub. As the students clapped, a girl shouted, "Can Chaser clean my room?"

I said, "That depends on how well you motivate her. At home, there may be twenty toys on the floor. But if I tell Chaser, 'Clean up and we will play Frisbee,' she will put the toys in the tub in two minutes and race to the front door. Do you have a Frisbee?"

"Not yet," she replied.

With laughter all around at that, Dave closed the meeting. If you're curious about the video, you can watch it at www.youtube.com /pilleyjw.

The demonstration gave me a big energy boost for working on the paper. In March, Alliston invited me to go with him to the Comparative Cognition Society's annual conference in Melbourne, Florida, where we would have an opportunity to report on Chaser's learning. The audience would be a tough-minded one, but receptive to the idea that a dog could reason. An increasing number of animal scientists around the world were moving away from Descartes's animals-are-just-meat-machines paradigm, and the conference attendees reflected that. I was too nervous to make the presentation, however, and I asked Alliston to do it. I was happy to be sitting in the third row listening to Chaser's story.

After Alliston's presentation several of the conference attendees asked tough questions, but in a friendly spirit that put me at ease. That evening at dinner I found myself sitting next to Clive Wynne. Alliston told me later that Clive asked to sit beside me because he was so intrigued by Chaser's learning. A transplanted Englishman, Clive was a full professor in the Psychology Department at the University of Florida's flagship campus in Gainesville, where he had his own Canine Cognition and Behavior Lab. He was also research director at Wolf Park, a research and public education facility in Battle Ground, Indiana, and editor in chief of the British journal *Behavioural Processes*.

Clive peppered me with questions about Chaser. He was fascinated to hear about the double-blind trial before Wofford's Psychology Club. And he was especially intrigued when I noted that Chaser performed language tasks for other trainers and questioners besides me.

"Suppose I were to come up to Spartanburg with a couple of students," he said. "Would Chaser be able to complete language tasks as you and Alliston describe if the students and I were the ones asking her to do them?"

"Sure," I said. "Come on ahead anytime."

A few weeks later, Clive drove up to Spartanburg with two of his postdoctoral students and research collaborators, Monique Udell and Nicole Dorey. Sally, Chaser, and I welcomed them, and we spent a little time getting to know one another. Chaser's way of doing that, of course, was to engage Clive, Monique, and Nicole in playing with her. Each of them was happy to oblige. And then I showed them the Rubbermaid containers with Chaser's toys so that they could conduct their own trials.

Clive, Monique, and Nicole randomly picked twenty objects, which we took to Wofford. In the same auditorium where Chaser had demonstrated her learning for the Psychology Kingdom, Clive, Monique, and Nicole arranged the toys in random order behind a large screen. Monique and Nicole took turns asking Chaser to retrieve the objects by name, while Clive evaluated the accuracy of each selection.

When Chaser quickly brought the first object correctly, Clive, Monique, and Nicole all exchanged a look that said, "Ummm, this could be interesting." With each retrieval, I could see their excitement growing.

Chaser brought the last of the objects out from behind the screen. She had not missed a single one. Clive turned to me and said, "If there's a trick to this, it's almost more impressive."

Sally and I were both bursting with pride. To see a leading canine cognition researcher and two of his best postdocs bowled over after testing Chaser's learning for themselves — well, I was on cloud nine.

Back at our house we all sat down together on the porch, and Sally and I answered our visitors' many questions about Chaser's training. For our part, we were fascinated to hear about their ongoing research projects. Their main focus was on how wolves and domestic dogs compare in their sensitivity to people's gazes and gestures. Discovering the differences and similarities between wolves and dogs in this regard would shed light on the evolution of dogs' social intelligence. Meanwhile Chaser livened things up by luring each of them into her favorite games. And then Sally and I got a shock.

Just before Clive, Monique, and Nicole left to drive back to Gainesville, Clive asked, "Would you consider letting us take Chaser to our lab to work with her there and then write up our own study?"

Seeing the negative reaction on Sally's face and mine, Clive said, "I don't mean now, of course. But would you consider it for sometime in the future?"

In order to be polite we said we would, and we thanked him for that extraordinary vote of confidence in Chaser's learning. Over the next couple of weeks I had several talks with Clive about the possibility of

Monique and Nicole's running a study with Chaser under his direction. The idea was tempting, because it would almost certainly assure that Chaser's learning would be shared with the scientific community at large through a peer-reviewed paper in a good journal. But there was really no way we could contemplate sending Chaser out of the family for weeks or months at a time.

Clive, Monique, and Nicole were all dog lovers. Sally and I had no doubt that they would take the best possible care of Chaser. But I finally explained to Clive that, as Sally and I always said, "Chaser is a member of our family." Clive understood. He had already invited Alliston and me to submit our paper for possible publication in *Behavioural Processes.* He now took the further step of telling Julia Cort, a producer at PBS's *Nova scienceNow,* about Chaser.

In late September, Clive provisionally accepted Alliston's and my paper. The peer reviewers for *Behavioural Processes* wanted to see a number of revisions and elaborations of our procedures, but they were all things I knew we could handle.

In October, Julia Cort brought a team — her assistant, a cameraman, and a soundman — to Spartanburg to shoot a segment with Chaser for a "How Smart Are Animals?" feature on *Nova scienceNow.* They arrived the evening before the planned shoot. Tall and slender with shoulder-length dark brown hair, Julia was as congenial in person as she had been on the phone and in her e-mails, and her team were all as nice as they could be.

Neil deGrasse Tyson, the host of *Nova scienceNow,* who would be in the planned segment with Chaser, arrived the next morning. Although Chaser's ability to connect with all sorts of people never ceased to amaze me, I had been a little concerned about how she would react to Julia, Neil, and the crew. If the meeting was in any way bumpy, there would not be much time to put everyone at ease. I needn't have worried, however. Chaser loved them all, especially Neil, who delighted her with his infectious enthusiasm and sense of fun.

Neil and Julia were as blown away by Chaser's language learning as Clive and his students had been a few months earlier. They told us that the feature wouldn't be ready to show on *Nova scienceNow* for at least a year, because of the other segments they were planning. They also warned us that they couldn't promise what would actually wind up in the finished program. But there was no mistaking their excitement over the footage they'd recorded with Chaser.

Their response energized me. Now I couldn't wait to finish revising the paper and get final approval of it by *Behavioural Processes*. The snag was that Alliston's commitments to SQAB really ramped up over the 2009–10 academic year. But finally in the summer of 2010 we were able to get together on the final changes to the paper. Alliston outdid himself in crunching the data on Chaser's learning and presenting it in dramatic figures and tables. Allston's contributions really drove home the stringency of my criteria for whether Chaser had learned a word, and the magnitude of the testing I had done.

With regard to Chaser's proper noun learning, for example, we now had 8,000 1-of-8 tests, 1,000 8-of-8 tests, 838 20-of-20 tests, and 145 formal blind 20-of-20 tests to report, as well as the double-blind demonstration at Wofford. Statistics for each kind of test showed that the p value, or probability-of-chance value, was always equal to or less than .004 — well below the point where Chaser's success could be attributed to chance. And the statistical evidence and blind and double-blind trials we presented on Chaser's other language learning — her combinatorial understanding of separate meanings for nouns and verbs, her grasp of common noun categories, and her ability to learn by exclusion — were equally significant.

Clive sent us the peer reviewers' final comments, with suggestions for a few minor additional revisions, in early September. He also suggested a bold title: "Border Collie Comprehends Object Names as Verbal Referents." It was not a title to excite the average person. But it was sure to grab the attention of any scientist with an interest in language learning by animals or children.

In November, *Behavioural Processes* formally accepted the paper, with e-publication to come the next month and print publication in February. And *Nova scienceNow*'s "How Smart Are Animals" program, with Chaser as the center of a "How Smart Are Dogs" segment, was scheduled to broadcast in February.

The stars seemed to be aligning favorably, giving us a chance to reveal to the world that dogs are smarter than we often think.

13

—

Going Viral

CHASER BECAME WORLD-FAMOUS before we knew it— literally.

It was early evening on Christmas Eve, 2010, and Sally and I were on the phone with Debbie in Brooklyn. Ten-year-old Aidan had just said good night after telling Sally and me what he was hoping Santa Claus would bring him for Christmas. Debbie said it was getting hard to hide presents from Aidan, but we were all happy that Christmas was still a magical time for him.

Debbie asked us what was new, and I remembered I had not yet told her that *Behavioural Processes* put Alliston's and my paper online on December 8. It had slipped my mind, and at any rate, the print publication wouldn't happen until February.

"Dad-d-d-d-d," Deb said in half humorous, half serious exasperation. "Why didn't you mention that before? In today's world the online publication is probably just as important."

I told her, "I don't know, honey. This is science, and the print edition will probably carry most of the weight."

Deb sighed and asked, "Can you just give me the exact title of the paper? I'll Google it." I did that, and we said good night.

Five minutes later the phone rang. It was Deb, practically breathless. "You won't believe what's happening!" she exclaimed. "When Jay and I Googled the name of the paper, all these links came up referring to Chaser as 'the world's smartest dog' and 'the dog with a thousand-word vocabulary.' Jay is scrolling through dozens and dozens of hits and he hasn't gotten anywhere near the end yet."

Deb asked Jay to pick up the other phone to tell us what he was seeing. He came on the line and said, "Chaser's gone viral, John."

Though I used the computer every day for e-mail, I wasn't Internet savvy and had to ask what that meant. Jay explained that news about Chaser was spreading on the Internet like a flu virus in a crowded room.

"Wow," I whispered.

Deb said, "Oh my god, Dad, this is crazy!" There was silence for a few seconds, and then she repeated, "Oh my god. Tell Mom to pick up the other phone, or put on your speakerphone."

As I switched on the speakerphone I asked, "What is it, Deb?"

She started speaking with an adrenaline rush: "I just had to read this a couple of times to make sure it was referring to Chaser. There is an article from the *New Scientist*"— she took a deep breath —"and here is how it opens: 'In the age-old war between cats and dogs, canines might just have struck THE KILLER BLOW! A Border collie called Chaser has been taught the names of one thousand and twenty-two items — more than any other animal'!"

"Wow!" I shouted. Startled, Chaser padded around the corner from where she'd been lying near the back door. In the winter she likes to feel the little draft of cold air slipping in under the door.

I quickly explained to Deb that Jessica Griggs of the *New Scientist*, a highly respected British popular science magazine, had called the morning of December 8. She'd read the *Behavioural Processes* article online a few hours earlier and had been waiting to call me because of the five-hour time difference between Great Britain and the eastern United States. She asked me a few questions and said she was going to talk to one or two experts to get an independent perspective on the paper. Alliston and I had been hoping she'd do a story in February when the print edition came out. I was so excited about this possibility that I made a deliberate effort to put it out of my mind, and I'd mentioned it only to Sally.

Jay came back on the phone and reported that the first hit was from two days before, on December 22, when the *New Scientist* website posted Jessica Griggs's article under the title "Border Collie Takes Record for Biggest Vocabulary." The next hit was from the following day, when the BBC's website ran a story headlined "Chaser the Border Collie 'Knows More Than 1,000 Words.'"

Jay said, "I'm jumping from site to site, but it looks like the BBC got its information from the *New Scientist.* And then after the BBC ran its item, the news exploded. There's stuff from all over the United States and the rest of the world: newspapers, radio and television stations, cable news networks, news websites, blogs, you name it. Wikipedia already has an entry about Chaser breaking the vocabulary record."

"You're famous, Chaser honey," Sally said, and Chaser trotted up to her to get a pet.

"The world's in awe of you, sweetie," I said, bending toward Chaser from my seat in the easy chair. Grinning and tail wagging, Chaser trotted over to me, and I rubbed the top of her head and behind her ears.

Encouraged by this attention and our obvious excitement, Chaser quickly got a blue racquetball and headed to the steps with it. At the foot of the steps, she turned and looked back at me expectantly.

"Chaser, you know this is not play time," I said gently.

Unwilling to take no for an answer, Chaser walked to the top of the stairs and turned around to look at me with the ball in her mouth.

"No, Chase," I said a little more firmly.

She sighed and lay down at the top of the stairs with the ball between her front paws.

I turned my full attention back to the phone conversation. Debbie was reading the brief but wonderful item on the BBC's website, ending with, "'It is thought the training may be the key to Chaser's apparently massive vocabulary.'"

"That's awesome," I said. Sally and I shared a kiss and beamed at each other, and we could hear the joy in Deb's and Jay's voices.

"You gotta look at these things online yourself, Dad," Debbie said. "The *New Scientist* has one of the videos of Chaser's testing that we put on YouTube for your and Alliston's paper. And practically all the other stories have embedded the same video or included a link to it. We went to YouTube fifteen minutes ago to see how many hits the video had there, and it only had forty. But we just checked again and there are already over a hundred."

"That's great," I said. "Maybe tomorrow there'll be a few hundred more."

Deb said she was e-mailing me the links to the *New Scientist* and BBC webpages, and we all said good night. Sally and I hugged and gave each other another kiss, and I turned to go upstairs to open Debbie's e-mail and print out the articles.

Before I put my foot on the bottom step of the stairs, Chaser nosed the blue racquetball over the top step. I chuckled and said, "Okay, Chaser. We'll play a little while. You should get to celebrate too."

For ten minutes or so Chaser and I played the game she had invented, taking turns bouncing the ball down the steps to each other. After I printed out the *New Scientist* and BBC articles, I took them downstairs to the living room to read and share with Sally.

Over the previous three years, as I struggled with writing the paper on Chaser's learning, Robin had repeatedly said, "You're going to be amazed by the reaction it gets." Robin has always been the most technologically savvy member of our family, but she had recently begun experimenting with living "off the grid." I couldn't wait to tell her how right her prediction was when we saw her on Christmas Day.

The opening line of the *New Scientist* article ("In the age-old war between cats and dogs . . .") was an impossible-to-ignore hook. The brief first paragraph captured the fact that Chaser's language learning included common noun concepts as well as her record-busting vocabulary of proper noun names. And it provocatively drew the connection with young children's language learning.

The rest of the article described the other findings in the *Behavioural Processes* paper, especially Chaser's ability to "infer the name of a new object" in learning by exclusion tests. There was a veiled reference to possible Clever Hans effects when Griggs quoted an expert in canine behavior and cognition: "'The experimenters did a lot of controls to exclude alternative explanations, although from my experience the results are simply too good,' says Ádám Miklósi, founder of the Family Dog Project at Eötvös Loránd University in Budapest, Hungary."

That was significant because Miklósi is perhaps the best-known European animal scientist specializing in research on dogs. His hint of skepticism didn't faze me. He was asking the same questions about our procedures that Alliston or I would have asked in his place. Miklósi saying that our results looked too good to be true but he couldn't find a flaw in our procedures was really a testament to Chaser's unprecedented learning and its potential significance. The article went on to say that Miklósi "thinks Chaser's intensive training explains the difference" between her results and those of other language-trained dogs.

It was all music to my ears. So was the brief article, more or less a capsule version of what was in the *New Scientist,* from the BBC's website.

While Sally and I read the articles, Debbie and Jay called back several times to report the growing number of Google and YouTube hits.

"It's like a pop song going number one with a bullet," Debbie said, as the hits for the YouTube video climbed from forty to more than a hundred thousand in a few hours' time.

Chaser caught the current of excitement in Sally's and my voices and body language, and she continued to want to play with her ball and other toys. Normally this was a quiet time, with me heading off to bed and Chaser settling down by Sally while she watched a television show or read a book before coming to bed herself. Sally and I happily indulged Chaser that night.

Debbie and Jay read a number of the YouTube viewers' comments to Sally and me on the phone. It was fascinating to hear how people were engaging with each other about Chaser and sharing stories about their own dogs. Quite a few viewers were debating the extent to which dogs could learn elements of human language, but always in a spirit of admiration for dogs' intelligence and problem-solving abilities. As a group the people commenting on Chaser's video repeatedly expressed an abiding love of dogs, gave thanks for the love dogs expressed in return, and stated the belief that their own dogs understood much of what was said to them.

The comments brought the same feelings to the surface in Sally and me. When we finally turned out the lights and wished each other and Chaser good night, we were aglow with gratitude and joy for our good fortune. I went to sleep thinking that Chaser's story could only have spread so far, so fast, because of the loving relationships that people experience with their dogs.

The next month was a whirlwind. The *New Scientist* article triggered interest in our work, and the BBC article sent Chaser's story racing around the globe. My contact information was on the *Behavioural Processes* paper, and every day my e-mail inbox teemed with fresh requests for media interviews and appearances. The phone started ringing on Christmas Day and didn't stop for weeks.

Chaser was the feel-good story of the holidays. When the holidays ended, the phone calls and e-mails didn't let up. Alliston and I were lucky that Wofford College's director of news services, Laura Corbin, took on the main burden of prioritizing media requests and scheduling interviews. Without her expert help, our heads would never have stopped spinning as we tried to sort things out. Or to borrow another analogy from Debbie, trying to keep up with the requests for interviews and information on Chaser was like trying to hit back balls from a hundred tennis machines at once, all of them set on rapid fire. As it was, Alliston and I each fielded hundreds of calls and did dozens of radio interviews. Alliston also did interviews in his fluent Spanish for media in Mexico and Spain. In late January, Laura Corbin tallied coverage for Chaser in more than forty-six languages around the world.

Deb was waiting in line at the supermarket one evening when she spotted the gossip tabloid the *National Examiner* and did a double take. Chaser was on the cover alongside Charlie Sheen and Brad Pitt. The copy next to Chaser's picture said, "DOG-GONE SMART! The world's brainiest pooch." Deb grabbed a copy to buy, telling the cashier, "I don't normally read this, but that's my dad's dog on the cover." With an I've-heard-that-one-before tone the cashier said, "Uh huh" and continued ringing up Deb's purchases.

In typical tabloid fashion the *National Examiner* had to knock someone, even in the canine world. The story said Rico only knew "a piddling" two hundred words and that "Chaser makes Rico look like a howling idiot." I found this upsetting, but Deb quickly put it into perspective for me: "You didn't say it, Dad. It's a tabloid and that's what they do — they slam people. It's hysterical that Chaser is on the cover with Brad Pitt."

Among the first journalists to call me for an interview was Annette Witheridge, the British correspondent in New York City for the *Daily Mail,* Britain's second-biggest newspaper. She asked if she could come to Spartanburg to see Chaser demonstrate her learning, and in the middle of the week between Christmas and New Year's, she arrived on our doorstep with the photographer Chris Bott. On January 1, 2011, the *Daily Mail* and the paper's website ran Annette's article, "Who's a VERY Clever Doggy! Prepare to Be Bow-Wowed as We Put Chaser, the World's Brainiest Dog, to the Test."

In a very down-to-earth way, Annette's article captured much of what I felt was important about Chaser's story, beginning with the fact that Chaser is a beloved member of the Pilley family and not simply a research subject. Annette really caught Chaser's personality, describing "her tail wagging so hard that half her body seems to be joining in" and evoking her social nature: "'Ooh, a new playmate; ooh, a new playmate,' she seems to be saying."

Annette's very personal response to meeting Chaser came to typify for me how people opened their hearts to Chaser and related her learning to memories of their own dogs. After asking me several questions, Annette shared her memories of Trixie, the collie mix her family had when she was a child. She proudly relayed that when anyone spelled W-A-L-K, Trixie knew what they meant. Hearing that made Sally giggle and she walked into the living room to say, "Chaser's exactly the same. When I say I'm off to pick up the M-A-I-L, she shoots out the door."

Perfectly on cue, Chaser sprang to her feet and went to the door, ready to help fetch the mail from our street-side mailbox. That's one of her important jobs, by the way. She insists on carrying a piece of mail into the house. It's the same when Sally comes home from the grocery store. Chaser insists on carrying something from the car to the kitchen. Once I gave her a banana to carry, which she did without leaving a mark on it. Since then I've been saying that we should try her out with an egg, but Sally says that's not practical and that Chaser's feelings would be hurt if she cracked the egg in her mouth and thought she'd let us down. Either that, or we'd need to buy a lot more eggs.

Alliston came over to participate in the interview, and Annette asked him to relate Chaser's abilities to the way children learn language. Alliston told her, "A child of two understands the phrase 'I love you.' I don't think Chaser would know that. By three, a child would say: 'Mummy, I love you' and know the meaning. Chaser couldn't do that."

I saw a small smile on Annette's face as Alliston said this, as if she felt sure Chaser knew the meaning of "I love you." In her story she quoted Alliston's remark, and then added, "As if to prove otherwise, Chaser . . . wags her tail and looks at me as if to say: 'Yes, I could.' And based on what I've seen today, I wouldn't put it past her."

The volume of e-mails and phone calls about Chaser increased even more after this piece appeared. One day Sally took a call from someone in Los Angeles who wanted Chaser to appear on a show there. We had decided that we weren't going to fly Chaser anywhere, because we didn't want her to have to travel apart from us in a plane's cargo hold. I heard Sally explain that, and focused my attention on answering some of the e-mail about the paper. Half an hour or so later I came downstairs for a drink of water, thankful that I hadn't heard the phone ringing and hoping things were starting to calm down. The calls had been coming in nonstop.

However, Sally was still chatting with the same person. It was a few more minutes before she hung up the phone and said, "What a nice young man." She was about to elaborate when the phone rang again.

It was Debbie, and Sally and I both got on the line. Deb had been trying to reach us and asked if the phone had been off the hook. Sally explained that she'd been talking to someone who wanted Chaser and me to appear on his television show in Los Angeles. Sally was getting frustrated with all the phone calls about Chaser, although she was always gracious when she answered the phone.

Sally said, "I explained that it was too much travel and we wouldn't put Chaser in a plane's cargo hold. Most people hang up when I say that, but he kept asking questions about Chaser and talking about his own dog. He really loves dogs. Finally after forty-five minutes I told him I needed to go."

Deb asked, "Did you get his name?"

Sally said, "It was Jimmy something."

"Jimmy Kimmel?"

"That's it! How did you know?"

Deb sighed and said, "Just a guess."

We especially enjoyed the visit to Spartanburg of a reporter and a photographer from the French magazine *Paris Match*. Having spent her junior year of college in Aix-en-Provence and being fluent in French, Robin was able to explain that *Paris Match* is like a combination of *People* and *Time* magazines. The reporter, Olivier O'Mahony, was as struck by Chaser's social nature as Annette Witheridge had been. His story described how Chaser brought him a ball as soon as she saw him and called her "the most sociable dog" he'd ever met: "In five seconds, I have made a new friend. In the world of humans, this takes more time."

After Chaser brought her ball over to make friends, I told her Olivier's name by pointing to him and saying, "Chaser, this is Olivier." As usual in such circumstances, just as when I show Chaser a new toy, she was lying on the floor and apparently looking the other way. Although I explained to Olivier that Chaser was giving me her ear and giving him a glance of her eye, he obviously had his doubts. So I asked him to go hide in the other room, and then said, "Chaser, find Olivier." He was astonished and delighted when she immediately went and nosed him out of his hiding place behind the couch.

Meanwhile, Sebastien Micke, Olivier's photographer colleague, was out in the yard in the snow, creating a display of Chaser's toys. Sebastien's pursuit of the perfect shot knew no bounds, and he lay full length in the snow to get it (see photo insert).

When the story on Chaser ran in the January 20–26, 2011, issue of *Paris Match,* it immediately followed one on John Travolta and his family. That tickled all of us, and Debbie said Chaser was now indisputably an A-list celebrity.

It was a very different situation when Nicholas Wade, the chief science editor for the *New York Times,* contacted Alliston and me. Like Jessica Griggs for the *New Scientist,* Nicholas Wade was all about the science. Many of his questions had to do with whether Chaser's results in various trials really avoided the Clever Hans effect, and he paid particular attention to the video, part of the online backup to the *Behavioural Processes* paper, that showed Chaser in a take-nose-paw test.

We video- and audio-recorded the test at Wofford College. The setup had me kneeling behind a screen that was thirty-nine inches high by four and a half feet wide. On a cloth in front of the screen were three of Chaser's toys: Lips, which is shaped like a pair of human lips; ABC, a cloth cube with those letters on its sides; and Lamb, a stuffed toy resembling a lamb. These were the objects I would ask Chaser to take in her mouth, nose, or paw. Sally sat at the side of the room more than fourteen feet away from the screen and just slightly behind me. Her job was to wave her hand when she saw Chaser take, nose, or paw an object. When Sally waved her hand, I said, "Good dog!" Sally then put the objects back in place in front of the screen while I rewarded Chaser with brief play with a ball, before we conducted another trial.

I had assigned a different number to each of the three objects and three commands, and used a random number table to pair commands and objects for fourteen trials. During each trial I couldn't see what Chaser was doing, and Chaser couldn't see me. Sally could see when Chaser did something with an object, but she was too far away to tell which object Chaser took, nosed, or pawed, or to cue Chaser's choices.

Afterward, three Wofford students independently watched the video with the sound turned off and wrote down Chaser's actions in all fourteen trials. Then they each independently watched the video with the sound turned on and assessed whether Chaser's actions matched my instructions. The raters unanimously agreed that Chaser got everything right.

Nicholas Wade interviewed Alliston and me separately about the video. His questions indicated that he had scrutinized it frame by frame, and that he had also discussed it and everything else in the paper with outside experts.

On the night of Monday, January 17, 2011, Nicholas Wade's article, "Sit. Stay. Parse. Good Girl!: Dog Might Provide Clues on How Language Is Acquired," appeared on the *New York Times* website. On Tuesday morning it was the lead story in the paper's weekly "Science Times" section.

The article gave a very thorough account of the *Behavioural Processes* paper and Chaser's language learning to that point in time. The opening paragraph put as big a smile on my face as the one in the *New Scientist*. It said:

> Chaser, a Border collie who lives in Spartanburg, S.C., has the largest vocabulary of any known dog. She knows 1,022 nouns, a record that displays unexpected depths of the canine mind and may help explain how children acquire language.

Wade discussed how the high working drive of Border collies helped explain why Chaser "proved to be a diligent student," and he compared Chaser's learning to Rico's. He drew out the implications of Chaser's abilities for understanding both nonhuman animal intelligence and children's language learning, "because children could be building on the same neural mechanisms." And he devoted six whole paragraphs to the danger of the Clever Hans effect and the safeguards against it in Chaser's language trials.

"Haunting almost every interaction between people and animals is the ghost of Clever Hans," Wade wrote. He quoted Alexandra Horowitz, scientist-author of *Inside of a Dog*, on Border collies' sensitivity to people's voices and attention cues as a possible source of Clever Hans effects. But he also noted that she said "the experimental design [in the *Behavioural Processes* paper] looks pretty good." Wade revealed that Horowitz had been one of the expert reviewers on the paper I submitted to *Science*, and I breathed another sigh of relief that this flawed first attempt to publish my findings with Chaser had been rejected.

Wade emphasized that Chaser's language trials followed the same rigorous procedures as in the Rico study at the Max Planck Institute for Evolutionary Anthropology in Germany. He quoted Juliane Kaminski, the lead author of that study, as saying, "I think the methodology the authors use here is absolutely sufficient to control for Clever Hans."

The specter of Clever Hans is one of the biggest thorns in the side of any researcher who wants to demonstrate learning by nonhuman animals. Although Jessica Griggs had referred to Clever Hans obliquely, her article did not mention his name or go into any detail. I was grateful to Wade for highlighting the most critical area that had to be controlled in my experiments with Chaser and my procedures for doing so.

I regularly taught my students about the horse Clever Hans. Owned by a high school math teacher named Wilhelm von Osten in early-twentieth-century Germany, Clever Hans could apparently understand spoken and written German, identify musical pitches, interpret clocks and calendars, and do arithmetic with both whole numbers and fractions. Hans amazed large crowds free of charge by counting off correct answers in sequences of numbers and letters with hoof taps.

In 1904, the German government's so-called Hans Commission decided that no deception was involved. But in 1907, the psychologist Oskar Pfungst showed that Hans was responding to the increase and release of tension in von Osten's posture and body language. When Hans saw von Osten tense up in anticipation of his answer, he started tapping his hoof. When he saw von Osten relax, he stopped tapping.

Von Osten had no idea he was cuing Hans, and Pfungst found that he himself and other questioners also produced involuntary cues in working with Hans. If questioners knew the answer, then Hans read their body language and almost always "answered" correctly. If questioners thought they knew the answers but were supplied with false information, then Hans again read their body language and "correctly" answered according to the questioners' false beliefs. But if the questioners simply did not know the answers, Hans only got an answer right now and then by pure chance.

The Clever Hans effect, as this involuntary cuing became known, is so pervasive and powerful that drug and bomb detection dogs may produce false positives in response to their handlers' body language. With their acute sensitivity to human body language, the dogs see that someone or something has aroused their handlers' suspicions and they react accordingly.

It's not hard to produce a Clever Hans effect intentionally, either. I described earlier how Robin did that in teaching Yasha to "count." From the start of my research with Chaser, I knew that other scientists would not accept my results unless I controlled absolutely for Clever Hans effects. In their paper in *Science,* the Rico researchers made a major point of their avoiding Clever Hans effects. In his critique of the

Rico study, the Yale psychologist Paul Bloom stressed the importance of their doing so successfully. And it was inevitable that peer reviewers would train a laser eye on whether I fell victim to the Clever Hans effect.

The need to avoid Clever Hans effects was not a stumbling block. It was a given of animal learning research that I had taken account of throughout my research in graduate school and at Wofford. I was glad to demonstrate in a crystal-clear manner that Chaser performed language tasks without any visual cues.

At the end of his article Wade came around, like the *New Scientist* and BBC articles, to how intensive training might explain Chaser's unprecedented results. Had I "lucked out in finding an Einstein of the [dog] species," or was I right in suggesting to him that "most Border collies, with special training, 'could be pretty close to where Chaser is'"? He gave the final word to Alexandra Horowitz, writing, "Dr. Horowitz agreed: 'It is not necessarily Chaser or Rico who is exceptional; it is the attention that is lavished on them,' she said."

Readers were quick to comment on the story on the *New York Times* website. After approximately 360 posts, comments were no longer being accepted. Scrolling through the posts on the site, I saw the same mix as for the YouTube videos of Chaser. There were some skeptics, but the overwhelming tone of the comments was one of celebration. It was wonderful to see people's enthusiasm for understanding their dogs better and enriching their relationships with them.

Sally and I especially enjoyed one reader's posting: "Cats can do this, too, they just don't want to." Another posting that made us laugh read: "I would like to propose a swap: two teenagers for Chaser. I'll throw in a chocolate cake and a couple twenties to sweeten the deal." Chaser has a temperamental moment now and then, but having raised two teenagers, we didn't want any part of that deal.

The print and online media coverage between Christmas 2010 and late January 2011, along with the outpouring of interest from the general public in website comments and blog posts, left me feeling profoundly grateful. The response to the paper in *Behavioural Processes* and to Chaser herself went beyond anything I had ever imagined. I hoped all the hoopla would encourage other researchers to try to duplicate my experiments with their own dogs. I also wanted dog owners to know how smart their dogs really are.

I had no desire to be famous, and I didn't want Chaser to remain in the media glare. Now that the first phase of her language learning had received a fair hearing, I wanted to return to our usual routine and see how much more she could achieve in the way of language and other conceptual learning.

But the *Nova scienceNow* program with Chaser was soon to air, and I was committed — very willingly — to doing publicity for it in New York. I hoped Chaser and I were both ready for prime time.

14

Chaser Takes a Bow

CHASER STOPPED IN a pool of light from a streetlamp. We were at the entrance to a public schoolyard a block and a half from Debbie and Jay's house in the Williamsburg neighborhood of Brooklyn, on the other side of the East River from Manhattan. The morning of Monday, February 7, was still dark, but Chaser and I kept to our early rising in New York City as well as Spartanburg.

The empty schoolyard was a favorite spot for Chaser and me, a fenced-in area where we could safely play with balls and Frisbees. But not this morning.

"Come on, Chase," I said. "Your paws have been getting all torn up playing in there, and we've got to follow the doctor's orders."

It had snowed every day for the past few days, and the sidewalks and streets—and the asphalt schoolyard—remained scattered with big crystals and clumps of road salt. The road salt was murder on Chaser's paws, and Dr. George Korin, the vet for Debbie's family's cats, had advised us to restrict her outdoor activity in these conditions.

Chaser didn't buy it. Her paws had healed up nicely over the past couple of days and she was restless to burn off some Border collie energy. She cocked her head at me and then pointed her nose into the schoolyard and woofed softly. This was the first time Sally and I had brought her to Brooklyn during the winter months, and she wasn't used to having so little physical activity.

"No, girl, we've got to take care of your paws," I said. "You don't want to go on national television in two days with bloody feet, do you?"

On Wednesday morning Alliston Reid, Chaser, and I were to appear live with Matt Lauer on NBC's *Today Show*. On Wednesday afternoon Chaser was going to record a segment with Diane Sawyer for that evening's *ABC World News* broadcast. *Nova scienceNow*'s outside publicist, Eileen Campion, who had worked in concert with Wofford College's Laura Corbin throughout January, had made the arrangements. The two appearances would publicize PBS's Wednesday-night broadcast of *Nova scienceNow*'s "How Smart Are Animals?" program, with Chaser featured in its "How Smart Are Dogs?" segment.

I gave Chaser a pet and we walked back to Debbie and Jay's row house a short distance from the Bedford Avenue subway stop. I loved the energy and variety of Williamsburg, and New York City in general, especially the opportunities to strike up conversations with people from all over the world. Although we both missed our usual outdoor activity in Spartanburg, Chaser also enjoyed the social opportunities in the Big Apple. She continued to look on every new acquaintance as a prospective playmate, and in Brooklyn there were many new people to meet every day, many new playmates to recruit for her games, no matter how briefly. It wasn't yet six a.m. and the streets were quiet. But they'd be bustling with people when we went out later in the day, and that would compensate Chaser a little bit socially for her forced inactivity.

Chaser enjoyed having new walking companions in the city too — Debbie and Jay often volunteered to walk Chaser when we came to visit. The first times either of them walked her on the street in Brooklyn, however, all heck broke loose. In Debbie's words the problem was that they didn't know how to "drive" Chaser: "You didn't tell us her commands, Dad. It's like trying to handle a Ferrari when you're used to a Volkswagen."

One afternoon Sally, Debbie, and I were walking home from the subway and saw a commotion on the sidewalk up ahead. When we got closer, we saw that the cause of it all was Jay trying to walk Chaser.

Tall and lean with chiseled features and shaggy dark brown hair that curls over his shirt collar, Jay ordinarily blended right into the Williamsburg scene. However, Chaser was walking backward in front of him with a cigarette butt in her mouth, spinning left or right every time he reached for it. This was drawing a mixture of smiles for her and glares for Jay from people who had to make room for them on the crowded sidewalk.

Jay was generously giving Chaser an airing because we were late getting back. His effort was all the more generous because he is highly allergic to dogs. And there Chaser was, giving him fits in return for his good deed.

Chaser likes to carry something in her mouth on walks and will pick up whatever she can find. In Spartanburg that's most often a pine cone. In Williamsburg pine cones are in short supply, but there are plenty of cigarette butts, and we often turned to find her with one dangling out of her mouth like a confirmed smoker. Jay was valiantly trying to get the darned cigarette butt out of her mouth because he knew we didn't like her carrying one around. But Chaser thought they were playing keep-away.

Throwing up his long arms at our approach, Jay said, "I didn't know I needed to read the user's manual!" A moment later, seeing Chaser walking perfectly at my side, he asked, "How did you do that?" I apologized for not having told Deb and Jay all of Chaser's commands and explained that I first said, "Time out," to signal that play was over for the moment; then, "Out" to get her to drop the cigarette butt; and finally, "Heel."

We were all laughing as we turned off crowded Bedford Avenue onto North Fourth Street, where the sidewalk was wide open. Even though Chaser is quite accomplished in her language training and has a huge repertoire of commands, she has the spirit of a puppy and will passionately pursue her desire to play at every opportunity.

Back in the house on the cold morning of February 7, Chaser and I walked up the stairs past the ground-floor apartment that Debbie and Jay rented to longtime tenants. The wide stair landing for the second floor had a water bowl for Chaser and several of her toys. When Jay was in the house, Chaser spent a good deal of time on the stair landing, where we frequently played the bouncing-ball-on-the-steps game she invented, or in the bedroom on the third floor that Sally and I shared with her on these visits.

We went through the door off the second-floor landing into the kitchen. No one else was up yet except the cats, Billy and Molly. We'd have breakfast with the family later, but in the meantime I got a snack for myself from the fridge and a couple of little dog biscuits for Chaser.

Billy, a gray tabby, swiped at Chaser's wagging tail when I gave her the biscuits. He likes to do that to show her whose house it is. When they first met, Billy had boldly walked up to Chaser, sniffed her nose, and then bopped it with his paw. That was exactly how Molly had greeted Billy on his joining the household four months earlier.

When Chaser eats her meals in the kitchen, Billy always sits right behind her and she growls at him now and then so he keeps his distance. Molly, a pastel gray calico, just watches from the sidelines. Her attitude seems to be, "Yeah, what Billy said!"

When no one else is around and she thinks she's in charge of the kitchen, Chaser tries to get her revenge by herding the cats. They foil her by jumping on the table. It's a stalemate then, as there sometimes is between Border collies and livestock on a farm or ranch, at least until the Border collies figure out how to solve the problem, with or without the shepherd's help. Chaser hasn't figured out how to herd the cats yet, however, and I don't think there's any way I can help her.

Chaser certainly has to adapt when we go to Brooklyn, but she seems to prefer it to being left behind, and we hate to leave her. Besides, visits to Brooklyn also mean lots of play time with Aidan. Chaser and Aidan became fast friends when she was a puppy and he was three years old. Aidan hadn't seen Chaser since the summer before in Spartanburg, so he was eager to get home from school every day to play with her.

Deb came into the kitchen. While she made coffee we spoke briefly about Wednesday's appearances on *Today* and *ABC World News*. Although we weren't sure exactly what was going to happen on *ABC World News,* I had no concerns about that appearance — even though I wouldn't be on camera with Chaser.

The general plan for *ABC World News* was for Chaser to demonstrate her language learning to Diane Sawyer while working with Neil deGrasse Tyson. Although they hadn't met since their first meeting during the *Nova scienceNow* shoot, fifteen months before, I knew Chaser would be glad to see Neil and comfortable working with him on camera. We were going to sort out what that involved and give Diane Sawyer and Chaser an opportunity to meet each other on Wednesday afternoon, before they recorded the segment for the evening's broadcast. I felt confident that Chaser's acute social intelligence and her language learning would enable her to handle whatever happened.

We'd been in Brooklyn for two weeks, and throughout that time I'd been trying to get a bead on how *Today* wanted to handle things. Alliston Reid was flying into New York the next afternoon. I thought Alliston, Chaser, and I should have an advance look at the set and run through the segment with the producer, and I hoped Chaser would have a chance to meet Matt Lauer before we went on the air live in front of millions of viewers. Like that of humans, a dog's behavior is always context sensitive and often situation specific. Change one detail of a process and you may change the outcome drastically.

From early on in Chaser's language learning, I envisioned public demonstrations in which Chaser worked in different settings besides our home or Wofford, and with different people besides myself. But until now Chaser had only solved language tasks in the same familiar surroundings where she was trained. I really wanted everything to go smoothly when Chaser was asked to demonstrate her vocabulary in the unfamiliar environment of the *Today* studio — and on live television, no less.

It was dark and the block was relatively deserted the next evening at six-thirty p.m., when a black SUV deposited Sally, Debbie, Chaser, and me, my arms wrapped around a plastic storage tub of toys, on West Forty-Eighth Street, directly in front of the stage door to NBC's Rockefeller Center studios. We rang the bell and were buzzed in by a security guard, who looked askance at the plastic tub I was carrying and even more so at Chaser.

"You can't bring that dog in here," the guard said gruffly.

"Oh, no, Charlie, they are with me," a lovely young woman with blond hair said, appearing as if out of nowhere. She turned to us and said brightly, "We are ready for you!" She looked down at her smartphone, which she held close to her in both hands, thumbs tapping away like mad. Without looking at us again, she turned and led us through a labyrinth of hallways and doorways, all while continuing to tap out messages on her phone and chat with us. A few minutes later we followed our multitasking guide out of the seemingly endless maze and onto *Today*'s set, where Alliston, Chaser, and I were going to meet the producer and run through the arrangements for our segment.

The empty set looked as though it was asleep, and my first impression was that it was much smaller than it seemed on television. The set areas that looked like broad, separate spaces on camera were tucked together only a few feet apart. Debbie, whose work in the music industry has included a number of television appearances, explained that this made it easier for the hosts to move from one spot to another in the course of the live broadcast.

Eileen Campion had told us that tomorrow's show would feature segments with First Lady Michelle Obama and the English actor-comedian Russell Brand, as well as Chaser. The lack of activity now reminded me of the night before Christmas, with packages beautifully wrapped under a tree gleaming with decorations and lights and everything in readiness for the activity and excitement of the morning.

I was trying to take in this strange new environment when a tall man in jeans and a flannel shirt, one of the stagehands, came up and silently took the tub of toys. With a smile he placed the tub on a nearby table, and I mentally kicked myself when I noticed it was crusted with dried mud from our backyard in Spartanburg.

A handsome young man walked onto the set, enthusiastically shook my hand, and identified himself as the producer of Chaser's segment. Talking as rapidly as he had on the phone the night before, when we'd arranged the run-through, he led me to a large red velvet curtain suspended from the rafters. On the floor below the curtain there was a carpet of bright green artificial grass. The young woman with the smartphone, the producer's assistant, followed along, still tapping away. A team of techs and stagehands also gathered around, waiting to hear what they needed to do for the segment.

Where was Chaser? We'd let her off the leash as soon as we entered the set area, and she'd dashed around a corner with Debbie in pursuit.

I pulled my thoughts back to the producer, who was still talking at a fast clip. He was explaining how the segment would unfold. When he paused briefly I said the red velvet curtain was going to be a stunning backdrop for Matt Lauer's interactions with Chaser.

The producer frowned and said no, the curtain was going to separate Chaser from Matt throughout the segment. Before I could respond, a loud bark reverberated through the set. Was there another dog here? I wondered. Chaser never barked full out like that. The producer continued explaining how the segment was going to go, using his hands now to emphasize his words. His assistant kept clicking away at her smartphone, but she flashed brief looks of concern that I wasn't following what her boss was saying.

I closed my eyes for a second and reminded myself to breathe. When I opened my eyes, Chaser skittered around a corner into view, bouncing a beach ball into the air with her nose. She jumped to bounce the ball into the air off her nose and keep it from hitting the floor, and barked between bounces. Debbie, who was trying to get ahold of her, later told me that she'd found the beach ball in the prop room. As Chaser came through the set area, she attracted the attention of all the techs and stagehands, who watched with wide eyes and open mouths. One of them exclaimed, "I've never seen a dog do that before! How did you teach her that?"

I hadn't taught her that. The women's volleyball team at Wofford had a few years before, when I was working with them on performance psychology. During breaks the members of the team enjoyed teaching Chaser to play volleyball with a very light ball that they sometimes used in their training. Chaser was now looking for someone to play volleyball with again. It was good that she was enjoying herself and discovering a new full-throated bark. But her somewhat dog-gone-wild behavior made me all the more concerned about her reactions to the intense atmosphere of a live television broadcast.

The producer's patience was evaporating. He wanted me to order Chaser to stand quietly on one side of the enormous red curtain so that we could begin rehearsing the segment.

I knew what he was proposing was not fair to Chaser and might result in a very disappointing segment for everyone, especially viewers. I tried to explain to him how both human and canine behaviors are context sensitive. Just as people behave differently in church and at a party, a dog will behave differently in different situations. Most dogs that know how to find the newspaper on the lawn and bring it into the house will be baffled if you ask them to find a newspaper inside the house and bring it to you. The request won't make any sense to them. "Chaser is very adaptable," I told the producer, "but this big curtain is too much to throw at her."

The producer said, "We've seen video of Chaser performing behind a screen, and I have been advised that we are to use this curtain."

The screens in the videos were there to ensure that Chaser could be guided only by the words she heard and not any physical cues, and they were more than big enough for that purpose. This huge curtain was going to isolate her in a way she'd never experienced, in a confusing new environment with lights, noise, and strangers she'd never encountered. I couldn't imagine putting her in that situation on live television in front of millions of people, and told the producer as much.

The producer said, "I thought we worked this out on the phone. I told you yesterday what we want you to do in the segment."

My jaw clenched involuntarily. I heard myself bellow, "I will not allow you to present my dog on live national television in a situation that sets her up to fail!"

The producer was taken aback. "I will have to check with my superiors about this!" he exclaimed.

If someone had dropped a pin then, everyone present would have heard it.

Finally I said, "If we can't find a compromise, I will take my dog home."

Fortunately Julia Cort, the ever-graceful executive producer of *Nova scienceNow,* appeared at my side and gently clasped my arm. She had arrived at the studio with Alliston and Eileen Campion's assistant, Vicky, while the segment producer and I were arguing. She said, "I'm sure we can work this out, John. No one wants you to put Chaser into a bad situation." Deb now stood next to us in silent support. Julia went over to the producer and they walked a few steps away. I heard Julia suggesting that Chaser would shine, and the audience would love it, if

Matt Lauer interacted with her as Neil deGrasse Tyson had on *Nova scienceNow*. Julia beckoned to me and said, "John, could you demonstrate that?"

I was still very rattled. A warm nose nuzzled my hand, and I looked down to see Chaser plastering her body to my knees and staring up at me with her huge brown eyes. She was imploring me to assure her that everything was okay, and she wouldn't leave my side until I did so. I looked up and saw Sally, Debbie, Alliston, Julia, and Vicky in a loose circle around Chaser and me. All the techs and stagehands were watching quietly. Even the producer's assistant had stopped tapping messages on her phone and was looking expectantly at me.

I knelt down to give Chaser a solid hug and whispered in her ear, "It's okay, girl." She immediately wiggled her body happily and nudged my cheek with her nose. I petted her some more, and stood up. Chaser dashed to the beach ball and nosed it up into the air for me to catch, which I did. I smiled at her, and in reply she grinned, tongue hanging out of her mouth, and wagged her tail.

I felt my tension melting away as I regained my confidence. I immediately knew how to get on the same page with our producer. I turned to him with an amicable grin and said gently, "Let me show you what Chaser can do."

We went back to the same area of the set. The enormous red curtain was gone, replaced by a backdrop banner with TODAY repeated over and over on it. The carpet of bright green artificial grass was still there. I grabbed a few toys from the plastic tub and tossed them onto the artificial grass. Chaser watched intently, ready to start herding her surrogate sheep in response to my words.

I knelt down and beckoned to the producer to kneel beside me. He shook his head no and politely said he was glad to stand back and observe. With equal politeness I said that he would gain a much better sense of Chaser's abilities if he got down on her level and gave her some commands himself.

The negotiations hung in the balance as the producer looked reluctantly at the floor. No one had ever completely resisted engaging with Chaser, but for a long moment I feared the producer would be the first. Finally he knelt beside me.

Thrilled to have another person at her level, Chaser made the next move. She grabbed Santie Claus in her mouth and, swinging her head straight up, she "tossed" it to the producer. He was taken by surprise and instinctively caught the doll in one hand.

I encouraged the producer to play a little bit with Chaser to make her comfortable with him. He tossed Santie Claus several feet away and said, "Fetch, Chaser." She bounded after the doll, picked it up in her mouth, and then returned to stand directly in front of the young man, locking her eyes on him. He leaned forward to take Santie Claus from her mouth, and she teasingly backed up while maintaining eye contact with him. He said, "Chaser," and patted his knee to indicate she should come to him. She stepped forward, never breaking eye contact, her ears up in full attention. Once again he leaned forward to take the toy from her mouth, and once again she teasingly backed up.

The producer cracked a small smile, and I could see the walls coming down as Chaser worked her wiles on him. "Congratulations!" I said. "You are now one of Chaser's many slaves."

The rest of the rehearsal went beautifully. As Chaser retrieved objects by name and took them in her mouth, nosed them, or pawed them on his commands, the producer saw that her language learning wasn't a stage trick performed in a rote way. Everyone was smiling as we said good night.

On the brief car ride back to Brooklyn, I tried not to think about the fact that Chaser's live national television debut was only a few hours away. I hoped I hadn't been too insistent with the producer, but a story of Debbie's had convinced me that I couldn't be careless about how the media presented Chaser.

As professional musicians, Debbie and Jay are both experienced performers. Yet they also both get nervous before any performance, big or small. When I asked why, Debbie told me about two friends, a pianist and a bass player, who meet for the first time in a long while. The pianist asks the bass player what he's been up to, and the bass player reports that he's recently completed a successful tour with a great band, recorded on a superstar's new album, and written the music for a hit movie. To each piece of good news the pianist says, "Yeah? I hadn't heard about that."

The bass player says, "The funny thing is that last week at a jam session I crashed and burned on a relatively simple tune."

The pianist says, "Yeah, I heard about that."

In other words, you are only as good as your last performance. I hoped Chaser's first live television gig wouldn't be her last.

In Brooklyn, Debbie unlocked the door to the house and went inside with Sally. Holding Chaser's leash loosely in my hand, I slipped through the door and let it close behind me. We were all trudging up the stairs when Sally said, "Where's Chaser?"

With an "Oh, gosh" I clambered back down the stairs. As I pulled the door open I heard Chaser's resonant new bark. She stood on the sidewalk wagging her tail and holding her leash in her mouth.

"I'm so sorry, girl," I said, reaching for her leash. Teasing me as she had the producer, she stepped back, dropped the leash, and gave another deep bark.

I laughed and said, "Come on now, Chaser. We've got to go to bed." She grabbed her leash in her mouth and backed up a little more.

At that moment, three young women rounded the corner and saw Chaser. The one in the middle exclaimed, "Too cute!" Her friends loudly agreed, and they all giggled and squealed more praise as Chaser wagged the whole back half of her body at them. She spotted a small stick on the sidewalk, picked it up in her mouth, and dropped it at the young women's feet, initiating a few minutes of play with them. I told the young women about Chaser's television debut the next morning, and they wished her luck as they waved goodbye. With that, Chaser was finally ready to go inside to bed.

The next morning, returning with Chaser from our normal predawn walk, I saw a black SUV with an NBC sign in the side window idling outside Debbie and Jay's house. A small knot tightened in my stomach.

Sally, Deb, and Aidan were all waiting in the kitchen, ready to go. Alliston, Julia, and Vicky were going to meet us at Rockefeller Center. Aidan was excited about getting to miss school and tag along to *Today* and *ABC World News*. With his Creamsicle cheeks, light brown hair, and creative imagination, Aidan reminds me of a young Tom Sawyer. The rest of the family says he looks like me when I was his age. His presence makes Chaser light up faster than I can say Frisbee, and she was excited about going on an outing with him.

We went down to get into the large SUV, which had two rows of back seats. Chaser hopped in right after Sally and sat on the seat beside her. In our own car she always gets to sit on the seat.

Wanting to be respectful of the driver and his vehicle's leather upholstery, I instructed Chaser to sit on the floor. It took two commands before she reluctantly left the seat for the floor. She sighed heavily as she lay down and rested her head on her front paws.

A few minutes later we were at NBC's Rockefeller Center studios. An NBC page whisked us to the *Today* green room. There was a little makeshift holding area for Chaser right outside, because animals were not allowed in the green room. When it was time to go to the set, Alliston, Chaser, and I followed the producer's assistant, taking a different route through the maze than the night before, until we reached the part of the set with the artificial grass and backdrop banner.

The assistant asked us to take our positions on the fake turf and wait for Matt Lauer. We had to be very quiet, because Meredith Vieira was conducting an interview only a few steps away. In contrast to the night before, the set was fully lit and crowded with people. Camera operators were catching every angle of the different set areas, as we could see on wall-mounted monitors.

In contrast to her exuberance the night before, Chaser was subdued, plainly feeling the impact of the bright lights and the activity all around. She lay down on the artificial grass as if she was bored, and she showed only mild interest when stagehands spread most of the twenty-five toys we'd brought in the plastic tub on the artificial grass and put the plastic tub to my left.

Matt Lauer walked briskly out from behind the backdrop banner. I felt his magnetism as he extended his hand toward Alliston and me with a friendly smile. And then he immediately dropped to one knee to introduce himself to Chaser. She rose to greet him, ears slightly back, tail a little low but still wagging. She glanced up at Matt for only a second, then looked at her toys.

"Do you want to play, Chaser?" Matt asked with a laugh. He grabbed a stuffed animal and tossed it in the air for her to catch. Her ears went up and her tail wagged vigorously as she caught it in her mouth. The entire studio seemed to breathe a collective sigh of relief, but I thought I might be projecting my own response. Matt played catch with Chaser for a minute, and only then did he stand up and quickly run through the segment with Alliston and me. Matt's consummate professionalism and unmistakable humanity, and Chaser's immediate positive response, made my concerns vanish as quickly as when Neil deGrasse Tyson had instantly made friends with her.

A good thing, too, because over the studio speakers we heard, "Chaser, stand by. And five, four, three, two, one."

We were live on air. Chaser lay in front of me with her head between her front paws and her ears up as Matt asked Alliston and me about the scientific significance of her learning. She knew intuitively that it was almost time for her to work. She was listening intently for the first words that would tell her what she needed to do.

As he had a minute before, Matt dropped to one knee facing Chaser, and she immediately sat up. Matt said, "Chaser, fetch Tennis. Fetch Tennis." Chaser stood on hearing her name, and on hearing "Tennis" walked toward two blue racquetballs a few feet apart from each other. When she was closer she saw that neither of them was Tennis. She came back toward Matt and he repeated encouragingly, "Chaser, fetch Tennis."

Chaser pricked up her ears when Matt spoke, looked toward him quizzically, and wheeled around to scan her toys. For whatever reason, probably the studio lighting, she didn't seem to see the yellow tennis ball that was Tennis amid the thick, bright green artificial grass. What humans see as yellow and green, dogs see as mostly yellow blurring into gray. In any case, Chaser hadn't gone near Tennis yet.

She stood in the center of the artificial grass, and her tail went down. My heart sank. But Matt wasn't giving up. As softly and urgently as David Johnson saying "You can do it," Matt whispered, "Chaser, fetch Tennis."

Chaser started to walk around to view her toys. She turned right wide enough so that a couple of steps finally brought her near Tennis, and she could identify it by its size and shape. She quickly picked it up in her mouth, triggering a little chorus of "awww"s from around the studio and a "Good girl!" from me as she took it to Matt.

When I saw the video later, I noted that it took sixteen seconds for Chaser to find Tennis and bring it to Matt. They were the longest sixteen seconds of my life.

Matt told Chaser to put the ball in the tub. She wanted to play with it first, following our pattern after she successfully completed a language task. But there wasn't time for play in the segment, and Chaser reluctantly tossed the ball into the tub, eliciting more "awww"s from around the studio.

Matt stroked Chaser on the head and told her, "Good girl!" And then he said, "Chaser, fetch Peppermint." Chaser went straight to Peppermint, a small white rubber barbell with pink and green stripes. She snatched it up in her mouth with a playful squeak, and teasingly squeaked it some more as she quickly brought it to Matt and dropped it in the tub.

Matt said, "Chaser, fetch SpongeBob." As she began to make a circuit of the more than twenty toys still scattered around, Matt repeated, "Fetch SpongeBob."

Chaser continued on her circuit. Her tail went up and her step quickened. But I didn't see SpongeBob.

"SpongeBob's not out there," I said apologetically.

Chaser kept heading across the stage and picked up something on the other side of Matt from me, as he said, "No, here it is right here."

Chaser brought the object to us, and I saw it was a Frisbee with SpongeBob's face on it.

"She sees better than I do," I said, feeling very relieved.

Chaser wanted to play with the Frisbee, but Matt told her to put it in the tub and she did so, resigned to its being out of commission for play.

Chaser was ready to fetch another toy and if possible play with it. However, there were only about forty seconds left of the segment's three-and-a-half-minute slot on the show. Matt stood up and asked what else we wanted to accomplish with Chaser. I put Alliston on the spot with a silly "Speak, Alliston." He good-naturedly replied, "Woof," and then explained that the next steps of language learning for Chaser would be working with syntax and how the order of words can affect meaning.

Meanwhile, Chaser had picked up KG, a toy barbell made out of purple rubber, and was squeaking it hopefully. She continued to squeak it as Alliston spoke, until I whispered firmly, "Out, out," meaning let go of the toy, and "Drop," meaning lie down. As she complied, like a child who knows that protest is no longer tolerated, I silently gave thanks that she had charmed her audience as usual.

The next moments flew by. Matt wrapped up the segment by telling viewers about that evening's *Nova scienceNow*. We stayed right where we were until the monitors switched to Russell Brand, sitting opposite Meredith Vieira. The show then cut to commercial, but not before Russell Brand wisecracked, in response to Meredith's saying it must be tough to follow a dog, "It's pretty easy, actually, because I could have put all those things in that tub. I was watching Chaser and I thought, 'That is *easy*.'"

Matt shook our hands and thanked us for coming and bringing Chaser, graciously posed for pictures, and then strode off to his next segment. Pumped up with excitement, I looked around for Chaser. Russell Brand was embracing her and posing for pictures with her and Aidan, until a producer rushed him off to join Matt Lauer in front of the crowd outside *Today*'s street-level studio.

Suddenly I was feeling as light as air, busting my buttons with pride over Chaser's national television debut. Matt Lauer could not have been more personable and amazing with Chaser, and the entire staff of *Today* had been great. Our segment's young producer had completely delivered and knocked the ball out of the park. What a great day, and it was only 8:45 a.m.

Here was Sally, beaming as she took my hand and squeezed it. She gave me a quick kiss and whispered, "Good job, Pill!" Julia Cort, also grinning from ear to ear over Chaser's performance, was asking Deb, who was snapping pictures, to be sure to send her copies. Soon we were all assembled, and Aidan proudly took charge of Chaser on her leash.

Plastic tub of Chaser's toys in hand, an NBC page guided us through the maze to the stage door on West Forty-Eighth Street. The same black SUV was waiting for us. The sidewalk was streaming with people hurrying to work, and Chaser basked in the smiles, compliments, and pets she elicited from the passing throng.

It was time to say goodbye to Alliston, who needed to catch a plane home. As his cab pulled away from the curb to take him to the airport, I thought about how lucky I was to have his friendship and his collaboration on the work with Chaser.

I turned back to the black SUV and saw that Deb was still trying to orchestrate getting everyone inside. Chaser kept jumping in and trying to claim a seat, and it took plenty of coaxing from Deb and me to convince her to get down. With great reluctance—and a dramatic sigh—Chaser plopped down on the floor behind the front passenger seat, and at last we were on our way.

We weren't scheduled to arrive at the ABC studios for the taping with Diane Sawyer until one-thirty. So that we could rest a little without having to fight morning rush hour and midday traffic going back and forth to Brooklyn, Julia had thoughtfully reserved a room for us at a dog-friendly luxury hotel in Chelsea.

The hotel was only twenty-five blocks away, but it took thirty minutes to get there because of the heavy traffic. Chaser ignored me and everyone else the whole way. I was a little frustrated with her sulking, but then I reflected that she had just brilliantly demonstrated her creative learning without receiving any of her usual play rewards. How many people watching her on *Today,* I wondered, had ever seen an animal demonstrate something without getting food rewards for completing a trick or a step in a routine? How many noticed that Chaser did everything she was asked without being given a single treat?

The hotel room had a king-size bed. As soon as Chaser saw it, she looked at me expectantly, wagging her tail.

"Go ahead, girl!" I said.

Chaser jumped up on the bed and spun around to face me with her ears pricked straight up at attention.

"Fast fan," I whispered. "Fast fan."

She hopped back a little, bowed her front legs, and shook her head back and forth with a low play-growl.

"Fast fan," I repeated, tossing her a toy misnamed "Ann." When Aidan saw the toy, he gravely informed me that it was Marvin the Martian from Looney Tunes. But I'd already written "Ann" on the toy in permanent ink, and that was the name Chaser learned.

Chaser has always loved jumping on our bed at home. She does this only when we invite her, never on her own. There is a ceiling fan above the foot of the bed, and she gets excited when we turn it on. She also likes the ceiling fans in the living room and upstairs, but the one in our bedroom is the most captivating, because it is right above the bouncy bed.

It was Robin who first noticed Chaser's fascination with the ceiling fans. And it was Robin who started the game by saying "fast fan" to attract Chaser's attention when she turned on one of them.

"Fast fan" can now mean play time anywhere, although most of all on a bed. But Chaser always looks up for a fan when she hears the phrase, because a whirring ceiling fan is the cherry on the ice cream sundae in her mind.

After Chaser and I played "fast fan" for a while, we all took a walk around the neighborhood. And then we rested for an hour before leaving for ABC in the black SUV. The now familiar seating to-do ensued until Chaser resigned herself to lying on the floor again.

"You'll only have to put up with this one more time today, Chaser," I said. But she was ignoring me and everyone else again.

At ABC's studios on West Sixty-Sixth Street on Manhattan's Upper West Side, we announced ourselves at the security desk. We figured we'd have to take Chaser upstairs to the *ABC World News* studio in a freight elevator.

Only a couple of minutes later, however, a camera crew showed up and began shooting footage. Our entrance was being recorded, as we only learned much later, for a promotional clip for that evening's news broadcast.

Chaser was intrigued by the cameraman, who stepped backward and forward and all around her for good angles, like a dancer. She didn't mind even when he brought the camera down right in front of her face for a tight close-up. She seemed to be thinking, "This is interesting. I want to know more about this game."

Or maybe she was just clicking into performance mode. In any case, she was alert and poised. Her posture was downright regal, I thought proudly.

The producer of the segment, a petite woman with brown hair, strode up and greeted us warmly. She led the way to an escalator to the mezzanine, where we were to take a regular elevator up to the newsroom and studio.

Escalators are not made for four-footed creatures. Chaser had certainly never been on one, and I thought I might have to pick her up in my arms. Hoping for the best, I plowed on with Chaser on the leash in front of me and the cameraman right behind. At the bottom step of the escalator I whispered, "Here we go, girl." Chaser hesitated almost imperceptibly, and then mounted the moving steps as if she'd done so countless times.

A few minutes later we got off the elevator on the newsroom floor and followed the producer's direction to proceed down a hallway, now with the cameraman in front of Chaser and me. Up ahead I could see people standing where the hallway opened into an aisle of cubicles on both sides. The closer we got, the more people stood up and came into the aisle, or stuck their heads out from behind their cubicles, to get a glimpse of Chaser.

Chaser responded with eagerness to meet them, and she was striding out ahead of me when Diane Sawyer came around a corner. She walked toward us down the hallway, as tall as the day is long and even more beautiful than on television. She extended her hand with a welcoming smile and said, "Hello, Professor. I'm Diane."

"Hey, Diane, it's so good to meet you," I said, shaking her hand.

"It's so great to meet you," she said. Bending to pet Chaser, she added, "And to meet the fabulous Chaser, too."

Chaser's regal poise was gone. She wiggled under Diane's affectionate pets like the puppy she has always remained inside.

After personally greeting everyone, including Aidan, Diane led us to a conference room, chatting with us and introducing us to a wide variety of colleagues along the way. It was impressive to see the positive spirit and serious sense of purpose of that group, and the mutual regard that Diane and her colleagues obviously shared.

The conference room had water and dog treats as well as refreshments for our little entourage. Our bucket of toys was already open on the floor, brought there by a stagehand. Before I knew it Chaser had her favorite blue racquetball in her mouth and was dropping it at Diane's feet.

"You really could try to play a little harder to get, Chaser," Deb said under her breath.

Delighted, Diane tossed the ball to Chaser. Chaser gave it back to her for another toss. And another. And another.

I gave Diane a brief demonstration of Chaser's response to herding commands. And Chaser pawed and nosed KG, the purple rubber barbell, and then took it in her mouth as I instructed her to do so in succession. The mischievous squeaks she gave KG when she picked it up with her mouth were her own way of punctuating the conversation.

Chaser was exhilarated in front of Diane and her colleagues. But she continued to listen and respond accurately to every command I gave her. I was so proud of her.

Just then Neil deGrasse Tyson loomed in the doorway. There is something about his presence, classic and imposing yet friendly, that reminds me of someone like Gregory Peck.

"Helloooo, Chaser," Neil said in his resonant baritone.

Chaser wiggled over to Neil and rubbed herself against his knees to receive his full-body pets. Finished cuddling, she showed him a toy they could play with, and he obliged while still managing to say hello to everyone in turn.

We brainstormed briefly about how the segment might go, but I saw that with two master communicators like Diane and Neil it was best just to set them going and let them develop the segment. What didn't work would be cut before it was broadcast.

The segment producer said we should go over to the set. Chaser and Aidan brought up the rear. Aidan was holding her leash in one hand, but he was resting the other hand on the top of her haunches, something that was possible only because they had been puppies together, so to speak, and had complete trust in each other. The hallway suddenly widened like a river delta, open offices on both sides emptying into the arena-like set of *ABC World News*. People were busy working throughout the high-tech space, oblivious to us.

We stopped there to wait for Diane, and Aidan became entranced by the spectacle all around him. He dropped Chaser's leash and walked forward a few steps, as if into a wondrous book or video game. Diane came onto the set, and Aidan self-consciously retreated to my side, not wanting to be anywhere he shouldn't. Seeing this, Diane beckoned to Aidan. She held out her hand to him and said, "Aidan, let me show you our prompter."

She took him by the hand and led him over behind her on-air desk, where she had him try reading the prompts off the monitors. Seeing Aidan relax and smile under her tutelage in Newscasting 101, I had no doubt that she would work the same magic with Chaser.

Diane and Neil briefly rehearsed their on-camera interactions, sitting on two sleek modern stools. They planned to talk about Chaser's achievements in between clips from *Nova scienceNow,* and to end the segment with a demonstration of her performing a language task in response to a command from Diane. I was relieved that I didn't have to appear on camera, but I still had a few butterflies on Chaser's behalf.

Diane wanted to see if Chaser would respond to her commands. Diane, Neil, and I pulled various toys out of the bucket, discussing their names and photogenic qualities. Very serious stuff.

We chose half a dozen toys, avoiding those Chaser retrieved on the *Today Show.* One of the toys Diane, Neil, and I picked out was her ABC stuffed block. When we'd put the toys into the plastic tub early that morning in Brooklyn, it was Aidan who'd said, "Here, Pop-Pop, let's take ABC." I was so nervous at the time that the coincidence of the toy's and the television network's names didn't occur to me. It was only now that I realized the significance of the toy's name for where we were—and Aidan's cleverness in suggesting we bring it.

Neil and I explained to Diane that all she had to do was tell Chaser what object to find. With the stools taken away, Diane and Neil knelt down on either side of the blue tub. Chaser popped to attention as Diane called her name. She tilted her head and locked her gaze on Diane's face.

Diane said, "Chaser, find Flipflopper."

Chaser immediately picked up Flipflopper, a Frisbee made of cloth, and brought it over. Diane threw her arms up into the air, then put her hands to the sides of her head and said, "Unbelievable!"

When it was time to record the segment, the lights went up with Diane and Neil on their stools. Chaser was at their feet among several toys, where I had given her a single command to stay. I hoped she would.

The moment Diane and Neil began to talk, they had Chaser's attention. After about thirty seconds, she lowered her head between her paws, but remained in a relaxed state of readiness. Every now and then she closed her eyes briefly as Neil spoke about the latest findings on animal intelligence and showed clips, mostly but not all including Chaser, from that night's *Nova scienceNow*.

It was a long wait for Chaser, who at one point raised her head and curled into a slight "C" to look over at me. I held my hand straight up and down, palm out, and mouthed, "Stay."

Almost three minutes into the three-minute-and-fifty-three-second segment Diane said, "Okay, I think we all want to see . . ."

Neil finished the sentence for her: ". . . a demonstration."

Chaser straightened herself out but remained lying down even as Diane and Neil abandoned their stools and knelt down beside her.

When Neil said, "Chaser," however, she immediately got to her feet. She turned around to face him and hear him say, "Find Goose." She pivoted around to scan the toys, took a quick step to find and pick up Goose, then pivoted back around and took a quick return step in front of Neil, squeaking the toy at him.

Diane and Neil both laughed, and Diane asked Neil, "Can I do this one?" She looked toward Chaser, who was looking all around the set and wagging her tail. Diane said, "Chaser," but she continued to look around the set. Diane said a little more firmly, "Chaser," and she turned to look at her with full attention. Diane said, "Find ABC."

Chaser sauntered over to the square cloth toy with the letters A, B, and C on its sides, grabbed it in her mouth, and turned to shake it at Diane, who said, "You did it! ABC! You did it!"

Diane looked at the camera and told viewers about that night's "truly remarkable" *Nova scienceNow.* Meanwhile, Chaser had plucked Goose out of the blue tub and was squeaking it, hoping to initiate a play session after successfully accomplishing her appointed language tasks. Spreading her arms in a gesture that embraced both Chaser and viewers, Diane concluded the segment — and assuming no breaking news required a change of plans, that night's broadcast — by saying, "And to you and your brilliant dogs at home, good night."

A few minutes later we all reconvened in the conference room. Unprompted, Diane and Neil both quickly wrote and signed personal notes to Aidan, who received them with a mixture of awe, gratitude, and delight. And then it was time for us to leave.

Going back out through the newsroom, Chaser was once again the happy center of attention. She wiggled and tail-wagged around the room to receive hugs and praise from as many people as possible.

The black SUV was waiting for us on West Sixty-Sixth Street. We all started piling into the vehicle, and before I knew it Chaser was sitting on the first row of seats behind the driver.

I gently said, "Get down, girl." She ignored me and looked in the other direction.

"Come on, Chase," I said. "You've got to get down."

She slunk down to the floor, and then climbed right back up on the seat.

"No, Chaser," I said. "On the floor!"

She grunted discontentedly, stood up on the seat, turned to face the window, and sat back down. I sat beside her and tried to nudge her down.

She wouldn't budge. So I gently tried to lift her off the seat toward the floor. I felt her move and thought she was going to get on the floor, but instead she threw all her weight against the back of the seat.

I couldn't bring myself to peel her off the seat after all she had done that day. She might have been oohed and aahed at and fussed over. But she had also put up with a lot of stress, especially being repeatedly denied play with her toys after she'd completed language tasks involving them. She had performed brilliantly. She had a right to be a cranky diva and want to sit on the seat as she normally did in our own car.

All the same, I felt helpless about there not being enough seats for everyone. I wondered if I should sit on the floor.

Julia graciously solved the problem, saying that she really should take a cab to her hotel and get some rest before a very early plane flight the next morning. I would happily have ridden on the floor in order for Julia to have dinner with us. But I didn't want to argue if she actually needed to prepare for her trip home from New York. We exchanged hugs, congratulations, and goodbyes, and then she departed in a cab.

Our stalwart driver took Sally, Debbie, me, and a sleeping Aidan and Chaser back to Brooklyn. We were all worn out, but it wasn't bedtime yet. First we had to watch the appearance on *Today,* which Jay had ready on the digital video recorder. When the segment reached the point where I said, "SpongeBob's not out there," I felt a rush of embarrassment over my mistake. But then it occurred to me that there could not be a better example of Chaser's performing a language task with no cuing from me. While I thought she was failing in front of my eyes, she was actually succeeding in front of millions of eyes much sharper than mine.

We were able to watch *ABC World News* at its normal East Coast time. The segment with Chaser closed the broadcast, Debbie quipped, like the headline act at an all-star concert. She knelt on the floor to give Chaser a hug and said, "Who's a rock star?"

I was elated about Chaser's performance on both programs. I was also thoroughly intrigued by the brief clip that *Today* aired from the "How Smart Are Dogs?" segment of that night's *Nova scienceNow,* and by the longer clips that *ABC World News* aired. There were still two hours before *Nova scienceNow.* In the meantime we ordered Indian food. And of course Chaser had to have a walk and some play time, as well as her own dinner.

We'd all been up since five a.m., and it was nine p.m. when we crowded into the little family room off the kitchen to watch *Nova scienceNow*'s "How Smart Are Animals?" program. Aidan curled up on the couch and quickly fell asleep. Chaser was oblivious to the television, but excited to be with everyone, including Jay. Although she respected her buddy Aidan's need for sleep, she interrupted the rest of us with her ball throughout the program.

After the excitement of the day, including seeing clips from the program on *Today* and *ABC World News*, I had a sense of anticlimax waiting for *Nova scienceNow* to start. But almost immediately I found myself watching with fascination, as if it portrayed a whole different family and their dog.

Written, produced, and directed by Julia Cort, the segment with Chaser, "How Smart Are Dogs?," was the first and longest part of the program (there were also segments on dolphins, octopi and cuttlefish, and Alex the parrot and his owner and trainer Irene Pepperberg).

All the footage with Chaser was shot at our home in the living room and backyard. I'd emphasized to every journalist I spoke to that Chaser was not just a research animal but a member of the Pilley family, and this footage really conveyed that beautifully. I was also happy to see that a few scenes in our backyard, showing Chaser going through her little agility course and catching a Frisbee, captured a flavor of our play together. Neil joined right into that play, and the program brought back how immediately and genuinely he forged a connection with Chaser.

With regard to Chaser's learning, the segment first showed all 1,022 of Chaser's named toys piled in the backyard. I took her into the house and upstairs to my study, with no view of the backyard, while Neil randomly picked nine toys. He put them behind the couch in the living room, and then called Chaser downstairs.

She retrieved the nine objects correctly in rapid succession. And her body language and expression as she listened to Neil displayed her utter confidence and complete command of her vocabulary. I felt grateful for how vividly the program captured that, as well as how it duplicated my experimental procedures, with Chaser unable to see anything in the living room while she was waiting upstairs in my study.

Following this came footage of the evolutionary anthropologist Brian Hare at Duke University and the psychologist Clive Wynne at Wolf Park in Indiana, with each of them offering perspective on the social intelligence of dogs. Then it was back to our living room in Spartanburg to see Chaser demonstrate her ability to learn by exclusion. Sitting on our somewhat frayed couch, Neil said, "Let's see what

[Chaser] does when we challenge her with a new toy she's never seen or heard the name of." Then he held up a toy and said, "I smuggled this into your house. It's a Charles Darwin doll." The doll looked very much like Charles Darwin in old age, bald on the top of his head but with a long fringe of white hair and a white beard.

While Chaser waited upstairs, Neil and I had put Darwin behind the couch with seven of Chaser's toys. Neil looked at the camera and said, "Okay, so [we] put seven toys behind the couch, plus Darwin. Chaser's never seen Darwin, hasn't even ever heard the name 'Darwin.' So we're going to see if she picks out Darwin by inference."

Neil called Chaser back, and the program showed her scurrying down the stairs, obviously eager to have more fun with him. First he asked her to find two of the named toys that were already in her flock of surrogate sheep, Sugar and Crawdad. With the same confidence as before she quickly retrieved them on command. It delighted her that Neil raised the pitch of his baritone voice to squeak "Excellent!" like a cartoon character after each retrieval.

Neil said, "Okay, here it comes, a name she's never heard before." Then he turned to Chaser and told her, "Find Darwin."

In contrast to the previous retrievals, Chaser cocked her head and looked intently at Neil when she heard "Darwin." After that slight pause she walked behind the couch. The cameras then showed her walking uncertainly around the toys behind the couch, and finally coming to a standstill directly over Darwin. Neil said in voice-over, "So, while searching for the other toys, Chaser knew exactly which one to pick up right away. Now, she seems to have to think about which one might be Darwin. . . . She takes so long, I call her back."

When Neil called Chaser, she came to the side of the couch and turned her head to look at him. He said, "Find Darwin." She looked at him for a second, then shifted her gaze to the front of the living room and then back to him, all with the same quizzical expression on her face.

"Find Darwin," Neil repeated one more time. Chaser tilted her head at him again, and then slowly walked behind the couch. She hesitated over the toys for several seconds, but then the camera showed her coming around the couch with Darwin in her mouth. Neil's delight at that delighted her, and she wagged her tail exultantly.

In voice-over Neil said, "I can't believe it. Chaser's never seen that doll before. Yet somehow, she made the connection that the word she'd never heard before went with the one toy she didn't recognize."

The segment ended soon after this, but we kept watching the program. I found the other segments fascinating too, and the profile of Irene Pepperberg's relationship with the parrot Alex, who had recently died, was very moving.

What a day it had been! Taking Chaser out for a last brief walk, I reflected on how lucky we were to have people like Matt Lauer, Diane Sawyer, and Neil deGrasse Tyson introduce her to America. In their own distinctive ways, they had each genuinely extended themselves to Chaser. In response, Chaser had adapted herself marvelously to each of them. Her interactions with Matt, Diane, and Neil demonstrated the abundant emotional and social intelligence that both sides must have for the dog-human relationship — and communication — to blossom to the full

A couple of days later, Sally, Chaser, and I returned to Spartanburg. It was time to get back to our normal lives. And Chaser and I needed to get on with her learning.

15

Chaser Goes to Washington

CHASER WAS A hit on television. But our next public challenge was demonstrating her learning for a potentially much more critical audience of scientists.

Back home in Spartanburg, Sally and I were glad to pick up our normal routines with Chaser. The media kept calling and e-mailing with requests for interviews and appearances. We politely declined them all, with the exception of a BBC *Super Smart Animals* program that gave us another opportunity to document Chaser's learning under rigorous conditions. Sally and Chaser resumed their daily walks with the Ya-Yas, and Chaser and I resumed our language learning research.

Chaser was glad to be home and doing our usual things, too. Nicholas Wade nailed it in his article on Chaser in the *New York Times* when he wrote, "Border collies are working dogs. They have a reputation for smartness, and they are highly motivated. They are bred to herd sheep indefatigably all day long. Absent that task, they must be given something else to do or they go stir crazy." For Chaser, the hours we spent each day working on language cognition tasks with her toys were the equivalent of time spent in the pasture herding sheep. The media appearances took Chaser away from the work and work-related play she needed for her quality of life. And they took me away from the work and play I needed for my quality of life, too.

In March the American Psychological Association, the main professional organization of scientists and clinicians in psychology, invited Alliston and me to give one of several plenary addresses at the association's 2011 annual convention in August in Washington, D.C. Alliston and I both felt honored by the invitation to present Chaser's learning at the APA convention, especially in a plenary address open to all attendees.

Unfortunately, Alliston had a conflicting commitment and couldn't accept the invitation. It would have been fun to share the moment with him. But I was delighted to accept the APA's invitation on my own, and excited about presenting Chaser's learning to an audience composed mostly of scientists.

It was going to take some work to prepare a talk that was equal to the occasion of a plenary address at the APA, however. Since I'd retired from teaching, scientific presentations had gone from slide projectors to digital "slide decks" assembled with PowerPoint or Keynote. On Debbie's advice I bought a MacBook Pro laptop, and she, Jay, and Robin all became my technical advisors and coaches in preparing a series of Keynote slides and video clips to illustrate my talk, which I eventually titled "Chaser and Her Toys: What a Dog Teaches Us About Cognition."

As I prepared the presentation, Chaser and I dived into extending her language understanding with regard to syntax (the grammatical structure of a sentence) and semantics (the meaning of the sentence). "Take ball to Frisbee" and "take Frisbee to ball" have the same syntax: a verb, a direct object, and an indirect object. Switching the places of "ball" and "Frisbee" gives them opposite meanings, however, and that's an example of semantics.

Chaser's performance in the take-nose-paw tests showed she could handle two elements of syntax, a verb and a direct object. Our next goal was to add a third element of syntax, an indirect object, as in "take ball to Frisbee." If Chaser could "take ball to Frisbee" and then "take Frisbee to ball," she would show an understanding of semantics as well as syntax.

I also wanted to see if Chaser's language and concept learning so far would enable her to match to sample and to learn by direct imitation of me. Matching to sample meant showing Chaser an object without naming it and asking her to find one just like it. The task may appear simple, but success requires drawing a mental inference or formulating an abstract concept, such as "he wants me to find what I now see."

Learning by direct imitation, performing whatever physical actions someone else performs, is also a much more complicated problem than it may seem. Imitation learning requires the mental awareness that the individual modeling the behavior wants you to copy it. This is an aspect of what is known as theory of mind: the awareness that another individual has a point of view different from your own. Believing that animals do not have a theory of mind, many scientists continue to insist that animals cannot imitate, and that what looks like imitation may only be instinctual behavior being triggered in different ways. Yet evidence is pouring in of imitation among species as diverse as bonobos and crows. I thought that if Chaser could learn new behaviors by imitating my actions, it might greatly accelerate her ability to learn complex new behaviors in the future.

August arrived on the heels of a brutally hot July with sweltering temperatures and no break in the weather in sight. We left for the APA convention very early on the first Thursday in August, and I waited until the last second to load Chaser into the car, after the trunk was packed and the inside was cooled off with the air conditioner on high. The trunk was full, because after spending Thursday night to Saturday in Washington, D.C., where Deb, Jay, and Aidan were meeting us, we were continuing north to spend a couple of weeks with them in Brooklyn.

Sally was already in the front passenger seat with her seat belt on when I brought Chaser out of the house. She stopped several feet away as I held the rear car door open. Experience told her the car was uncomfortably hot.

"Hoop, Chase," I said, giving her the usual signal to jump up onto something, whether it was a bed, the couch, or the back seat of our car. No matter how I repeated or elaborated that command, however, she refused to budge.

"Come on, girl," I said. "We're going on a trip."

She stood her ground, eyes meeting mine with her ears back. The look on her face seemed to plead, "We don't really need to go on a trip, do we? Let's stay home where it's nice and cool."

I asked her again to get in the car, putting a little more force into my words. She didn't move an inch.

I went over to her and said, "You want me to help you?" With that I awkwardly lifted and shoved her into the back seat, inadvertently twisting her hind legs in the process. She quickly righted herself and plopped down on the opposite side of the seat.

Three hours later we stopped to rest and play with Chaser. "Hoop, Chaser," I said when it was time to get back in the car. But again she ignored that signal and my repetitions of it.

The last thing I wanted to do was to try to push and lift her into the car again. But finally I walked over to her saying, "You want me to help —" Before I could finish saying "help" she jumped into the car onto the back seat, where she turned around and looked at me warily.

When we were back on the highway, it slowly dawned on me that my clumsy attempt to get her into the car earlier that morning might have planted the idea in her mind that "You want me to help you?" meant "I'm gonna twist your hind legs now." When we stopped again, we went through the same rigamarole until I said, "You want me to help —," at which point she hopped into the car. I poured on the praise and petted her to help build a positive association on top of her aversive memory of being pushed uncomfortably into the car.

After a little more than six hours of driving we arrived at the Westin Washington, D.C., City Center on M Street, about a mile from the Walter E. Washington Convention Center, where the APA was holding its convention. I was eager to get over to the convention center and get a flavor of what was going on. I had never joined the APA during my career as a psychology professor — they made me a member for the year because of my plenary address — and I had never been to the annual convention. I also wanted to make sure that I had everything I needed to plug my computer into the convention center's audio-visual system and show my Keynote presentation.

Sally and I decided we should walk from the hotel, thinking it would be a chance for all of us to stretch our legs after being cooped up in the car for so long. After twenty minutes and several wrong turns, we were melting with sweat and barely halfway there. My laptop felt heavy in my briefcase, which fortunately had a shoulder strap. The temperature was in the upper nineties with oppressive humidity. Spartanburg is in the foothills of the Appalachians, and the hot weather we were now experiencing was even worse than what we'd been having at home.

Forty-five minutes after setting out, we finally reached the immense convention center. We practically seeped into the entrance area, blessing the cool air. A security guard immediately said, "You can't bring that dog in here."

I was trying to formulate a parched response when we heard a woman say, "That's okay." It was Candy Won, the APA's meeting director and the person who had conveyed the invitation to Alliston and me. She happened to be walking by when she spotted Chaser, and she was coming over to greet us when the security guard spoke. Candy graciously waited while Sally and I drank deep at the water fountains and filled up a portable bowl for Chaser to lap up every drop of. Then she walked us through the registration process and arranged for a volunteer to guide us around the convention center.

In addition to my plenary address, Candy and I had arranged for Chaser and me to give four thirty-minute demonstrations. Two demonstrations were scheduled for after the address on Friday afternoon, and two for Saturday morning. At my request the volunteer guide took us to see the rooms where I would give the address and the demonstrations. An audio-visual technician was meeting us in the room for the plenary address, so that I could make sure my Keynote presentation was ready to go. As we made our way through the convention throngs, people frequently stopped us to exclaim over Chaser and pet her, to her great delight, and say they were coming to the address and demonstrations.

The lecture hall looked imposing to me from the entrance. A four-foot-high stage ran across the front of the room with steps at either end. A long table on the right side of the stage had room for eight panelists all facing the audience. At the left was a podium, and above the main part of the stage hung a huge screen to display a speaker's slides or video.

I asked our guide how many people the hall seated. "A little over four hundred," he said. "The halls for the plenary addresses are all the same size. But as you've just seen, people are really buzzing about yours. Everyone's curious about Chaser."

I focused on the task at hand and forged through a quick rehearsal pretty easily. The tech was a great help, and it reassured me to see my Keynote slides and video clips pop onto the hall's big screen. I was feeling good about my preparation as Sally, Chaser, and I slowly walked back to the hotel in the oppressive heat. It didn't occur to us to try to take a cab with Chaser, but at least we knew the way now.

Our timing was perfect, however. We arrived just as Deb, Jay, and Aidan drove up to the hotel entrance. After they checked into the hotel and got settled in their room, it was time for dinner. I begged off joining everyone and asked them to bring me something back. I wanted to rehearse my presentation a few more times, even though I felt pretty good about it. Deb had shared her and Jay's experience that the technical aspects of an event that are outside their control as musicians, the sound and the lights and so on, easily and commonly go wrong. The key to dealing with technical glitches when, rather than if, they happened was to be on top of your material and confident enough to navigate the inevitable bumps and bobbles. I wanted to heed Deb's words of wisdom.

Fifteen minutes later I felt like one of my students who meant to stay in the dorm and study but kept thinking that all his or her friends were out having fun. I'd had about enough of rehearsing and wished I had gone to dinner with everyone.

Sally called just then and asked what I wanted her to bring me back for dinner. They were only a couple of blocks away, and I hurriedly put on my shoes to go join them. Chaser stood up on the bed, where I'd invited her to get, and looked expectantly at me with her head tilted sideways, hoping that I was taking her with me.

I told her I'd be back. She responded by wagging her tail, jumping down off the bed toward me, and again locking eyes with me and giving me her "can't I go, please?" look. Feeling a twinge of guilt, I said, "No, girl, Pop-Pop will be back."

Chaser turned around and jumped back on the bed, flopping herself down in an excellent approximation of a teenage girl's pout. She heaved a dramatic diva's sigh as she lowered her head to her paws. Standing at the open door, I repeated, "I'll be back, girl." She shot me another glance, almost rolling her eyes, and seemed to breathe another deep sigh as she settled down for a comfortable doze.

When we returned about an hour later, Chaser gave soft little "Hurry up" woofs as I fumbled with the electronic key. Once I had the door open I had to find her leash, but I saw that the message light was blinking on the phone. There were two messages requesting an interview from a journalist named Sharon Jayson, who had left two messages earlier in the day and whose number I had already scrawled on a copy of *USA Today*.

As Deb and I searched for Chaser's leash, I mentioned that the same woman had called four times. Chaser wanted to get outside and was following at my heels as I hunted for the missing leash.

"Well, who is she?" Deb asked, looking under the bed.

Fishing through Chaser's toys and growing more irritated about her missing leash, I said, "I don't know, I don't want to be speaking to reporters."

Deb calmly replied, "Well, give me her number and I'll call her back." I passed Deb the copy of *USA Today* on which I'd written Sharon Jayson's phone number.

Deb hung up the phone just as I found Chaser's leash in my briefcase. Chaser immediately grabbed a Frisbee and headed for the door as soon as I pulled out her leash. "Did you talk to Sharon Jayson?" I asked Deb.

"Yep," Deb said. "Her deadline is right after your talk, so she wanted to check some facts first."

"Well, it's good you called her back. It could have been awkward to run into her tomorrow. Did she say what publication?"

"Yep," Deb said. "*USA Today*."

The next day came quickly. At five a.m. Chaser and I left Sally sleeping in our room to go for a walk and some play. The previous evening we'd played with a Frisbee for a while in a small courtyard next to the hotel and in a triangular green space with benches on the other side of M Street. But the weather remained stifling hot.

I closed the hotel room door softly and turned around to walk to the elevators with Chaser. But she had dropped her small cream-colored polyester Frisbee, named Snow, by our room door and trotted down the hall to wait for a throw.

It hit me that this was a precisely analogous situation to our walking outside onto our front porch at home in Spartanburg. Chaser's modus operandi there is to drop her ball or Frisbee on the porch and then proceed out onto the front lawn to await a throw.

I smiled to myself at the cleverness of all dogs, and not least of all Chaser, when it comes to inveigling people into playing and interacting with them. I picked up Snow, intending to carry it with me to the elevator, but when I had the Frisbee in my hand I couldn't resist sailing it down the hall to Chaser. She caught it in the air, brought it within a few feet of me, and then raced back down the hall.

Oh, well, I thought, *why not have a few throws here before we go out into the heat and humidity?* Chaser seemed to be in no rush to get outside to do her business. And although she'd discovered a full-throated bark at the *Today Show* and employed it every once in a while out of excitement or frustration, she remained a rather quiet dog who preferred to vocalize with soft woofs. Her woofs only turned to barks if the woofs failed to draw our attention.

We played in the thickly carpeted hall for about twenty minutes, and then Chaser went to the elevators and woofed, softly, that it was time to go outside. Reentering the hotel ten minutes later, we both were grateful for the air conditioning. Getting out of the elevator on our floor, I turned to head to our room, but Chaser was pointing herself in the other direction.

"It's this way, girl," I said. "Or did I make a mistake?"

I checked the room number signs on the hotel corridor wall. No, I was right and Chaser was wrong, and that puzzled me.

"Come on, Chase," I said. But she still looked reluctantly in the other direction. Only then I remembered that Debbie, Jay, and Aidan's room was down that way.

"Do you want to see Aidan?" I asked. Chaser and Aidan had had a great time playing together the night before.

Chaser wagged her tail vigorously on hearing Aidan's name. Mystery solved: Chaser and I were both right. But it was still only six a.m.

"We'll see Aidan later. Come on now," I told Chaser, but she kept looking down the corridor. So I did what I always did when I wanted to end a play session without disappointing Chaser, and said, "Let's go see Nanny." At the sound of "Nanny"— it could just as well have been "Sally"— Chaser wheeled around and began trotting briskly toward our room. Following her, I had to laugh at myself, recalling my once telling Wayne West and his fellow Border collie trainers that their dogs didn't understand personal names.

Inside our room Sally was stirring, and she sleepily asked how our walk was. When I described Chaser's discovery that the hallway was excellent for Frisbee play, Sally smiled at first. But then she said, "You can't keep doing that. You're going to wake people up." Discretion is the better part of valor, and I didn't disagree with her.

We still had all of the morning and early afternoon to wait through until my three p.m. presentation. But I realized I was looking forward to the lecture and even more so to the demonstrations. I was eager to hear what other psychologists had to say when they observed Chaser's learning up close.

Finally it was time to go. We opted for cabs so we didn't have to worry about parking, and much to my relief, the driver of our cab had no problem letting Chaser ride on the back seat between Sally and me. The convention center security guards had apparently all been advised about Chaser, and she and I waltzed right in. I felt myself being swept up in the energy and activity surrounding us. Off we trekked to our presentation, up the escalator, which Chaser handled with aplomb, and through wide lounge areas with charging stations for electronic devices. It reminded me of changing planes in a big airport. We had to stop a few times along the way so Chaser could respond enthusiastically to convention attendees who asked to meet and pet her, and who told us they were looking forward to my talk and demonstrations. At the entrance to the lecture hall there was now a two-by-three-foot sign on an easel with a sign-up sheet for the demonstrations after the talk, and people were waiting to put down their names.

The hall was already half filled with attendees. We still had twenty minutes to go, and heads turned as people realized Chaser had entered the room. A wave of "awww"s followed us as we approached the stage, with Chaser walking steadily beside Aidan. The room started to buzz a bit more as I climbed the stairs and headed toward the podium, while Chaser stood at the front of the auditorium awaiting a command. Two photographers with cameras on tripods were on either side of the room, and a cameraman with a large handheld video camera was in the center aisle. The three of them expertly faded into the background as they focused their cameras on Chaser.

Jay helped me plug my laptop into the hall's projection system. As we were cuing up my presentation, I noticed a smiling woman introducing herself to Deb, Sally, Aidan, and Chaser. I turned my attention back to my laptop and clicked through the first few of my images. They appeared on the big screen in good order, and I clicked back to the first slide, with the title of my talk on it.

Whew! Everything seemed to be set. Jay left the stage to sit beside Deb at the near end of the front row. I glanced around. All the seats looked full, and people were standing along the back and sides of the hall. My nerves ratcheted up a few more notches. I stared down blindly at my notes and checked my watch. Still a few minutes to go.

Sometimes you think you're feeling one thing when you're really feeling another. Excitement and fear can trigger the same physical responses in your body, accelerating your pulse and breathing rate with a surge of adrenaline. *I'm excited,* I told myself. *Just go with it.*

Deb brought the smiling woman to the podium to meet me. She was Dr. Nancy Dess, a professor of psychology at Occidental College in Los Angeles, and she was going to introduce my talk. Deb returned to her seat in the front row, and Dr. Dess and I discussed where we should sit before she went to the microphone and whether Chaser should come on stage.

The audience instinctively quieted down as Dr. Dess approached the podium. *Here we go,* I thought. Dr. Dess gave me a lovely introduction. As she finished the audience applauded, and they seemed to clap louder as I walked to the microphone.

I began speaking, and Deb rushed to the front of the stage to tell me I had to speak directly into the microphone. I squared up to the microphone and asked if they could hear me in the back.

"Louder!" the whole audience seemed to say.

I leaned closer to the microphone, and asked, "Can you hear me now?"

"Yes!" the audience said. I clicked on my Keynote program, glanced up at the big screen, and saw that my first picture of Chaser was upside down. *Well, John Pilley, what do you do now?* I thought. After diligent preparation to avoid this precise predicament, I stood in front of four hundred faces not knowing what magic button to push. I stared at the screen on my laptop while everyone else stared at me.

A man shouted out in the audience. I ignored him and continued staring at the screen on my laptop as if my eyes could twist Chaser right side up. The man in the audience became more insistent. Realizing that he was not heckling me but giving computer directions, I uttered an inner "Hallelujah!" This man was going to be my salvation. I felt my anxiety disappear as I surrendered to the situation. Falling back on my years of experience as a professor as well as a minister, I faced the crowd head on, leaned forward to the microphone, and said, "Help!"

Laughs rippled through the audience, and the helpful fellow shouted, "Press control-alt-R." I repeated his instructions aloud with a laughing "okay" and pressed control-alt-R. The picture of Chaser popped upright on my laptop and the big screen above my head, and the audience applauded and laughed as I joked about my lack of technical skills.

First I clicked through a brisk slide show of Chaser as a puppy. The audience's reaction showed that psychologists were not immune to the power of puppies to warm our emotions. And then I showed a slide summarizing the four experiments that Alliston and I reported in our *Behavioural Processes* paper, followed by a video clip from *Nova scienceNow* of Neil deGrasse Tyson's testing Chaser in learning by exclusion with the Charles Darwin doll.

When the images onscreen shifted from the *Nova scienceNow* clip to my own clumsy video, I quipped that no one would be calling me to produce their movies anytime soon. More laughter broke out in the audience, and that relaxed me completely. I was able to forget about my notes and speak to the audience conversationally, and they gave me their undivided attention as I described the experiments Alliston and I had reported in *Behavioural Processes*.

In conclusion I said, "Helen Keller awakened to the meaning and power of words as Anne Sullivan drew *w-a-t-e-r* on one of her hands while water from a pump flowed over her other hand. In that moment Helen realized that Anne was tracing the letters of a word — and what that word meant. That simple but crucial insight opened the eyes of Helen Keller's mind to a lifetime of learning."

I paused for several seconds and then said, "Chaser's moment of awakening came in her first year of life, when she discovered that objects have names and learned the cues that enable her to match a name to an object. That simple but crucial insight opened the eyes of Chaser's mind to the ABCs of language."

There was absolute silence as I drew a long breath. I opened the floor to questions and hands shot up all over the hall while the rest of the audience applauded loudly.

It fascinated me that, allowing for their professional expertise and comfort with technical vocabulary, this audience of psychologists asked basically the same questions as the media and the general public. They wanted to know if I thought Chaser was unique, how her learning compared to that of one- to three-year-old toddlers, and what advice I had for teaching dogs words and training dogs in general.

As I always did, I said that I thought other Border collies could likely achieve similar results with similar training, but that other dogs might prove just as able to learn, and that linguistics researchers would have to answer the question about toddlers. To teach a dog the names of objects, I advised simple repetition of "This is . . ." associations, using play with the object to give it and its name value in the dog's mind. The key to all training and teaching of dogs, I said, was play based in a relationship of mutual trust and affection.

Hands kept shooting up to ask variations of these questions, often delivered with an anecdote about the questioner's own dog. As the clock ticked toward four p.m., ten minutes past the allotted time for my address and audience questions, Dr. Dess came to the podium and announced that we had to stop. But she urged everyone to sign up for the demonstrations later that afternoon and the next morning.

A standing ovation followed, but Dr. Dess's announcement didn't really end the session. It just brought people surging to the front of the stage to ask questions, offer congratulations, tell me about their own dogs, and most of all get close to Chaser. She was on the auditorium floor in front of the stage, tail wagging with joy as she received affectionate pets, hugs, and even belly rubs from her new fans. Members of the audience were snapping photos of her with their phones like paparazzi pursuing the hottest celebrity of the moment.

The first member of the audience to introduce herself to me was Sharon Jayson of *USA Today*. She told me to look for her story in the evening edition, and then rushed off to make her deadline. A neuroscientist told me about working with her dog on modulating barks in an effort to approximate word sounds, and asked what I thought the chances were that a dog could learn to speak words. I said I wondered if the anatomy of a dog's larynx, vocal chords, and jaw would allow that, but I hoped she'd let me know about her progress.

Chaser's fans would have kept us there longer, but we eventually had to excuse ourselves so that I could have a few minutes' break before the first demonstration at five p.m. Candy Won and I had agreed on a limit of fifty people per demonstration session, and the designated rooms were a good size for that many people and perhaps a few more. When we got to the room where the first two demonstrations were to take place, there were at least sixty people there, and others squeezed in as the demonstration got under way. I recognized quite a few faces from the audience at my talk.

There was a stage at one end with a podium and a long table for panelists at other events in that space, but I thought it would be more fun to do the demonstration in the center of the room. I said hello to everyone and sat down on the floor with Chaser and Aidan, who grinned and blushed self-consciously when I introduced him as my assistant. Aidan's first job was getting some of Chaser's toys out of the tub of toys we'd brought and spreading them around on the floor. Chaser then demonstrated her ability to retrieve a variety of her toys by their proper noun names, and she showed her combinatorial understanding in a take-nose-paw test.

I asked if anyone had objects they'd liked to see Chaser learn the name of by exclusion. While people dug into their pockets, backpacks, and bags, Aidan took Chaser outside so she wouldn't see what the group picked and named for her to find. The group offered up everything from coin purses to small flashlights and umbrellas. A yellow nylon wallet with a Velcro closure caught my eye. Chaser had a purse among her toys, but no wallet. Although none of her toys was named Wallet, she had certainly heard me say that word around the house many times. But "Velcro" was definitely an unfamiliar word for her, so I suggested we use the nylon wallet and call it Velcro.

First I asked the man who owned the wallet if he wanted to empty it. "I trust Chaser," he said, drawing a big laugh. I put the wallet on the floor with seven of Chaser's toys, and then Deb went out to get Aidan and Chaser.

First Chaser found two of her toys when I asked for them by name, demonstrating that she did not have an overriding preference for picking the novel object. When she heard "Chaser! Find Velcro," she paused just as she had when Neil deGrasse Tyson had asked her to find the Charles Darwin doll on *Nova scienceNow*. I repeated, "Find Velcro," and there was a hush in the room as she carefully examined the objects on the floor for several seconds and then picked up the yellow wallet in her mouth. It was always fun to hear the "awww"s when people saw Chaser do exclusion learning with an object they supplied. It was much more impressive than any video for them to see that happen right before their eyes.

I then briefly described Chaser's progress in learning by imitation, matching to sample, and comprehending a three-elements-of-syntax sentence and its semantic reversal. Everyone wanted to see Chaser demonstrate all three, but we were running long. I said that although we were in the early stages of training in all three areas, Chaser was doing best with the imitation learning so far, and that, assuming they didn't mind watching me lose my dignity, they might find that the most interesting. After saying, "Chaser, watch Pop-Pop. Do what I do," I lay prone on the floor, rolled over, and got up on all fours. And then I said, "Now you do it." However, I'm afraid the audience was as impressed with my agility at eighty-three as they were with Chaser's execution of the same movements after me.

They liked it better when I did a figure eight around two chairs, walked to the end of the room and up onto the stage, circled the long table, and came back to the starting point. This was obviously a behavior that Chaser would be extremely unlikely to emit on her own, and everyone was stunned to see her do it. They were even more stunned when Chaser did it all once more when I said, "Again." I explained that Chaser also understood "faster," "slower," and "reverse," which she demonstrated when I asked her to imitate a figure eight around two different chairs.

The demonstration was scheduled to last thirty minutes. It ran to forty minutes, right up against the start of the second demonstration, which was just as successful and ran just as long as the first.

We piled into a couple of cabs to go back to the hotel. Recalling our car seat struggles in New York, I was glad once again that Washington, D.C., cabdrivers didn't mind Chaser's riding up on the seat.

A copy of *USA Today* with Sharon Jayson's story was waiting for us at the hotel:

RESEARCH DOG REVEALS CANINES' COGNITIVE POTENTIAL

WASHINGTON — *Don't underestimate the mind of a mutt — at least when the canine in question is one of the smartest breeds and her trainer has painstakingly taught her to identify more than 1,000 objects.*

The dog of this day is Chaser, a 7-year-old Border collie who can also distinguish between nouns and verbs. She's a featured player at the American Psychological Association's annual meeting here, where her owner and trainer — retired psychology professor John Pilley — today let Chaser bask in the attention at the first of several weekend demonstrations planned.

From the moment this popular pooch entered the room, she was clearly the star, with cameras flashing and people maneuvering for shots of the brainy dog. . . .

We opened a bottle of wine in our room and toasted Chaser. She grabbed her Frisbee Snow in her mouth and shook it as she jumped on the bed. (Sorry, Westin.) And then we all got our supper and spent the evening unwinding together.

On Saturday morning Chaser and I played Frisbee again in the hallway, although we moved down from our door a bit in case Sally heard us. I don't believe we disturbed anyone else either, but I can only swear to the fact that no one opened a door to complain.

Our demonstrations that morning were again standing-room-only events, and after lunch we were all ready to drive to New York in our two cars. Debbie was going to go with Chaser and me, and Sally was going to ride with Aidan and Jay in their car. When I opened the rear car door for Chaser, she stood stock-still just as she had on Friday morning outside our house in Spartanburg and during our two stops along the road.

But as soon as I began to say "You want me to help you," she leaped into the car onto the back seat and turned around to look at me. I was already praising her warmly, and I saw that her wariness was diminishing. In fact, she was looking quite pleased with herself. The expression on her face seemed to say, "You think I'm gonna let you lift me into the car? Ha-ha, fooled you again!" It was like her game of dropping a pine cone in front of Sally or me on a walk, and then racing to grab it before we could pick it up.

Ever since then, all I have to do to get Chaser in the car is to start saying "You want me to help you?" I wouldn't say she's forgotten entirely that my helping her into the car that very hot morning in Spartanburg was a little uncomfortable. But we've layered pleasant associations on top of it, so that we can both feel good about how she gets into the car now.

The journey with Chaser always has the same number one priority: fun. Sally and I find fun with her every day, a routine that never becomes stale.

16

Expanding the Conversation

I<small>T'S A ROUTINE</small> of surprises. Most mornings around five a.m., Chaser and I are on the ten-minute drive to Wofford. As we have since Chaser was a puppy, we're going to hold our first training-and-play session of the day in the exercise center. But we don't know exactly what will happen while we're there. When we turn in to the campus, Chaser stands up on the car seat, tongue lolling out and tail wagging eagerly.

The surprises begin as soon as we enter the Richardson Physical Activities Building. Chaser wheels to face me, and I bounce her one of her blue racquetballs.

Chaser catches Blue and immediately drops it at my feet. I kick the ball down the hall and she darts ahead to snag it on the bounce, carom, or roll, twisting and turning to keep up with its crazy course. She drops it at my feet as I walk down the hall, and I give it another kick.

A few more paces and the exercise center opens up on our right, with weight machines, treadmills, and other equipment. There's usually an early bird or two working out, and others come in as the morning advances. If Chaser knows the people, she offers her ball for a kick or throw. With new people, she tries to make friends with the same approach.

I give her time only for brief hellos and pets at this point, and we go into a large room off the main exercise area. The room is about twenty-five feet by forty feet, with two glass-wall-backed racquetball courts on either side. In addition to being an entry and waiting area for the courts, it has mats and exercise balls for stretching. On the opposite side of the room is a doorway to a hallway that we also use for training. There's clear glass around the doors at either end of the room, and it's a wonderful space for us because it's self-contained but doesn't feel closed in. Sometimes half a dozen people or more will stand on the other side of the glass and observe us. I always wave to them. If we are near the end of the session I may invite them in, which Chaser loves.

Chaser and I begin with a fast-paced five minutes of play with several named toys to rehearse her language learning. I randomly ask her to fetch, shake, catch, toss, take in her mouth, nose, paw, and herd one toy after another. And I kick and throw the ball for her to capture as it ricochets around the room.

From there we work on new or recent learning. We alternate brief trials in the current lesson with brief periods of play. The play provides more opportunities to rehearse her language learning as she chases and gathers her toys while I verbalize what she is doing or what I want her to do next. After we've worked on the current language task for ten to fifteen minutes, we follow the same fast-paced rhythm as we move on to match-to-sample and imitation learning.

Play with Chaser's toys continues to enhance the value of the toys and the language tasks and games involving them. What is more important is that the play refreshes us both, that it continues to confirm and deepen our relationship as fellow creatures. That is really what enables her language learning to progress.

We know that the progress of toddlers' language learning depends on their language-based relationships with parents, siblings, and others. Toddlers whose parents speak to them on all sorts of topics throughout the day advance much faster and farther in language learning than toddlers whose parents rarely speak to them except to scold them or tell them to do something. Somehow, the mystery of what happens in children's minds when they first acquire a language and the relationships they have with other people are interdependent.

As children acquire language, they also acquire a sense of themselves as social beings. As they learn the meaning of words, they also learn that other people have unique points of view, thoughts, and feelings. As they develop an implicit understanding of grammar, they also develop the ability to infer how cues such as pointing, facial expression, and tone of voice indicate the meaning of words. No one knows precisely how these things reinforce each other, but language is inherently a social activity.

If I am less than enthusiastic in a language trial or the play following it, Chaser's engagement fades. She delights in pleasing me, and if my enthusiasm flags, she naturally thinks of something she's sure we'll enjoy. That's when she entices me to play with a ball or a Frisbee. In the end it is always the quality of our relationship that determines the quality of the learning.

We repeat the same language, match-to-sample, and imitation learning that we do at Wofford in the rest of the day's training sessions at home. But our early-morning visits to Wofford stand out for both of us because of the opportunities to engage with new and old friends among Wofford students, faculty, staff, and administrators.

One day we were working on a language task when half the women's varsity soccer team came by with their coach. Within seconds Chaser was getting belly rubs and other pets from the players. Smiling at this scene, the coach asked if we could do a brief demonstration.

I called Chaser to my side and told her, "Chaser, watch ball. Watch ball." I rolled a racquetball onto the floor and said, "Go out, Chase. Go out, go out." She raced to circle around behind the ball.

"Chaser, come by. Come by, come by." She wheeled clockwise around the ball.

"Way to me, way to me." She wheeled counterclockwise around the ball.

"Walk up, walk up." She approached the ball.

"Drop." She instantly went to her belly.

"One, two, three, take!" She sprang forward and grabbed the ball in her mouth.

"That'll do." She ran to me with the ball.

"Good girl! Good dog!" I said, stroking her side.

The players crowded around us, and I stepped back to allow Chaser to wiggle and wag and flop on the floor for their pets, embraces, and cooing words. She was never sharper than in performing for this impromptu admiring audience, and she was in heaven with their praise and attention.

Unpredictable rewards for behavior motivate more powerfully than predictable ones. When Chaser and I went back to our language training, she was extra sharp at that too, because it was so closely associated in her mind with the team's arrival and her interaction with them.

Moments of discovery in language learning don't seem to loom up in Chaser's conscious mind the same way as discovering how to connect with new people. But like toddlers unconsciously understanding words as symbols, Chaser experiences unpredictable breakthroughs in language learning too. These are moments when the challenge of a language puzzle turns into a confident prelude to play.

The puzzle we were working through after the APA was understanding a sentence with three elements of grammar. I was excited, as always, to help Chaser experience more of the mystery of language acquisition, even if it was only a rudimentary language.

My training method continued to focus on what the eminent animal scientist John Staddon calls creative learning rather than rote learning. In rote learning the goal is to teach a predetermined response. In creative learning the goal is to stimulate and support spontaneous responses to solve a challenge.

Shepherds train Border collies using creative learning. Once Border collies have learned a few basic obedience and herding commands, they are literally turned loose with the sheep. Their knowledge of how to herd sheep then develops as they spontaneously behave in instinctual ways and the shepherd positively or negatively reinforces their choices of what to do.

As I mentioned in chapter 7, John Staddon has likened the way a wise teacher encourages a student's individual interests in learning to the way the early-nineteenth-century Scottish "shepherd poet" James Hogg trained his Border collie Sirrah. Hogg described this training as putting Sirrah into situations in which "he would try everywhere . . . till he found out what I wanted him to do." In the process Sirrah demonstrated "a great share of reasoning."

The result of Hogg's open-ended training was that Sirrah was later able to gather and safeguard seven hundred lambs that were scattered from their pens during a storm in the middle of the night, with no assistance or direction from Hogg or anyone else.

Staddon sums up creative learning for animals by saying that "it means creating an environment in which the animal's natural propensities (which, in an intelligent animal, go far beyond reflex response) can flower to their full extent." Once we've created a positive way for an intelligent animal to tackle a challenging problem, we can let a natural creative learning process occur. Our job is to watch for the spontaneous problem-solving efforts that have the most promise, and reinforce them positively. As the educator Maria Montessori put it, "Never help a child with a task at which he feels he can succeed."

That is what I've tried to do with Chaser. I've presented her with challenges just outside her grasp and encouraged her spontaneous efforts to solve them. Along the way I've seen her acquire greater and greater implicit understanding of words, making it possible to present her with even tougher challenges. Learning truly builds on learning.

Chaser and I started working on sentences with three elements of grammar late in 2010. As I mentioned, some researchers question whether a dog's correct responses to two-word sentences, such as "fetch sock," demonstrate combinatorial understanding of two elements of grammar, a verb and a direct object, or understanding of a single element, a "fetch-the-sock" fusion. Chaser's take-nose-paw tests demonstrated her independent understanding of verbs and direct objects. She was ready to tackle a third element of grammar, an indirect object.

The question was how to add the indirect object. What action should I ask her to perform, and how should I structure the sentence to communicate that?

Early in her first year of life, Chaser learned that *to* means "go to" or "take to." When Chaser already had an object in her mouth, Sally and I frequently told her, "To Pop-Pop," "To Nanny," "To living room," "To front porch," and so on. Likewise, early in her training, Chaser learned that *take* meant "hold the designated object in your mouth," independent of any future action with the object.

"Take ball to Frisbee" did not work well, however, because of Chaser's tendency to focus most intently on the last thing she hears. For Border collies the shepherd's last word or phrase contains the essence of the command. With "take ball to Frisbee" she tended to pick up the Frisbee rather than the ball.

Not long before the address and demonstrations at the APA, it occurred to me that the two halves of the sentence might be clearer for Chaser, and easy enough for me to say, with a structure of "to Frisbee take ball." The sequence of grammatical elements then became *prepositional object, verb, direct object*.

This sentence structure made the last thing I said the first thing Chaser needed, and wanted, to do: pick up an object in her mouth. The structure also followed the format of herding commands, in which the last words a dog hears represent what he or she has to do next.

The first times I told Chaser "to Frisbee take ball" and "to ball take Frisbee," with only those two objects on the floor, there seemed to be less hesitation in her responses. But I wasn't sure if the difference was real. Over the following few days, however, it was obvious that her confidence was increasing.

Alliston told me, "You're using the same syntax structure as in Spanish." He is fluent in the language from his years of teaching in Mexico. It tickled me that I had stumbled onto a sentence format used in a real human language.

Finding that a Spanish-style sentence clicked for Chaser was a neat aha moment for both of us. The experiment could proceed and the play could continue.

I selected a hundred of Chaser's 1,022 named toys as a training group. Working with random pairs of toys, I frequently reversed the roles — prepositional object and direct object — that two toys had. I also reversed their positions on the floor, sometimes putting the direct object on the right and sometimes on the left.

At first I pointed to each toy as I spoke its name. Over time I phased out the pointing. I also complicated the task by putting two possible prepositional objects and two possible direct objects on the floor. Later I put the two prepositional objects and the two direct objects in different rooms.

An early training trial began with two toys five to ten feet apart on the floor. With Chaser close in front of me and both of us facing the toys, I said, for example, "Chaser! To Sugar take Decoy." She had to pick up Decoy in her mouth, carry and drop it near Sugar, and then take Sugar in her mouth. Completing the trial brought Chaser praise and play with Sugar.

If Chaser headed toward the wrong object or got hung up in indecision at any point, I recalled her without correction and repeated, "To Sugar take Decoy" or whatever the initial instruction was. As always I wanted Chaser to start off with errorless learning and gain confidence with each success.

As we worked on "to Sugar take Decoy" and "to Decoy take Sugar" type sentences, there was less need to recall Chaser and repeat the instructions. If she hesitated, I said, "Do it, girl. Do it!" Enthusiastic encouragement usually emboldened her to make a choice, and usually it was the right one. If she kept hesitating I recalled her without correction and gave her the "to A take B" instruction again. Her comfort with that gave me the impression that hesitating became her way of saying, "Could you please repeat that sentence?"

It was fascinating to see Chaser try various strategies for responding to sentences like "to Santie Claus take Flipflopper" and "to Flipflopper take Santie Claus." After being reinforced for picking the toy on the right a few times in a row, for example, Chaser apparently formed the hypothesis that I always wanted the toy on that side. Gradually she realized that this strategy was unreliable.

It also emerged that how I said "to A take B" mattered quite a lot to Chaser's comprehension. She was best able to process the words and hold them in working memory when I slowly but emphatically said, "to A, take B," with a definite pause between "to A" and "take B" and rising energy as I completed the sentence. The more enthusiastic and encouraging I made "take B" sound, the better.

Early in 2012 the journal *Learning and Motivation* invited Alliston and me to submit a paper on Chaser for a special 2013 issue on animal learning. Alliston didn't have time, but he suggested I do it on my own. I decided to report my three-elements-of-grammar experiments, assuming the results were statistically significant.

In addition to my training-and-play sessions with Chaser, Sally and

I incorporated three-elements-of-grammar sentences into our interactions with her, including our nighttime routine. She plays and snuggles on the bed with us, lying in between Sally and me and being petted by both of us, until we say, "That's enough, Chase." At that point she curls up on the end of the bed for a while, before moving off to the living room couch and other favored sleeping spots. "To Nanny, take Sugar," "to Pop-Pop, take Decoy," and "to bedroom, take Crawdad" closed the day's training at home, as "to ball, take Frisbee" and "to Frisbee, take ball" trials began it at Wofford.

By fall 2012, Chaser became virtually perfect in responding to three-elements-of-grammar sentences in informal trials and play with Sally and me. She could complete the reversal trial (to B, take A) even if it did not follow the initial trial (to A, take B) until after intervening trials with other object pairs. It was another instance of relationships powering Chaser's learning.

Seeing Chaser's accuracy exceed 90 percent in informal trials and play convinced me that she was ready to work with other testers in formal trials under double-blind conditions. The double-blind tests would provide results for the paper for *Learning and Motivation*.

The tests were blind in that the testers, Wofford student volunteers furnished with instruction sheets with the commands for each trial, did not know the names of Chaser's toys. So there was no chance of Clever Hans cues. The tests were double blind in that other students, who were not present during the tests, evaluated audio-video record ings of them. During the tests, I operated a camcorder to capture audio and video.

We did the double-blind tests in two experimental scenarios with random pairings, varied placement of toys, and initial and reversal trials for each pairing. In experiment 1, using toys from the training group, Chaser got twenty-five of thirty-two right, or 78 percent. In experiment 2, using toys we had not used in training the sentences, she got eighteen of twenty-four right, or 75 percent. In both cases the probability that her correct answers resulted purely from chance was less than one in a thousand. This showed that her understanding of this kind of sentence could not be attributed to chance factors.

When Alliston watched the video of experiment 2, he noticed that

Chaser looked briefly at the prepositional object and the direct object as she heard their names, and then went straight to the direct object. This suggested that Chaser's brain might not be processing a mental image of the direct object into working memory, because she did not take her eyes off it after hearing its name.

To test this possibility, I devised experiment 3 using six random object pairs. For each pair there was an initial trial and a reversal trial, with two possible indirect objects an inch apart on the living room floor and two possible direct objects an inch apart on a pillow at the head of Sally's and my bed.

Standing in front of Chaser as she lay on the foot of the bed facing me, so that she could not see the toys on the pillow, I gave her the "to A, take B" command. When she stood up and turned around to select one of the direct objects, I could not give her a visual cue because the objects were so close together. Having made her choice of a direct object, Chaser raced with it into the living room and raced back with her choice of a prepositional object, which we played with before starting the next trial.

Chaser got all the trials in experiment 3 correct, with a less than one in a million probability that her choices resulted from random chance. Together with the first two experiments, this indicates that her brain does process a mental image of the direct object into working memory when she hears its name.

The three experiments also demonstrate both that Chaser securely lodged the names of all the toys and the meanings of "to" and "take" into long-term memory, and that she successfully held two toy names, "to," and "take" in short-term working memory while she made a semantic judgment about which toy to pick up first. The results provide strong evidence that Chaser can understand the syntax and semantics of sentences with three elements of grammar. And in showing that she can combine two cognitive abilities, long-term memory and working memory, to solve language tasks, the results raise the bar in terms of expectations for a dog's language learning.

Chaser's results suggest the possibility that she experiences a "phonological loop" when she completes "to A, take B" tasks. A phonological loop occurs when people repeat the directions for a task silently in their heads until the task is done. Researchers with access to brain scanning equipment might track a canine phonological loop by comparing activity patterns in a dog's brain to those in human brains when people silently repeat a sentence to themselves.

The dolphins Phoenix and Akeakamai respectively got 62.3 percent and 56 percent of their three elements of grammar sentences correct. In her equivalent trials Chaser got 78 percent and 75 percent correct. Taken together, these results indicate that an animal's brain can in some way operate like a language-learning toddler's brain and display implicit understanding of rudimentary grammar.

No one teaches little children the rules of grammar — they may never learn the formal rules of grammar — but somehow they acquire an implicit understanding of it. From the first babbling to the first sentences, from "Mama" and "Dada" to "More juice" to "I love you" to "Tell me a story," they listen and respond in ways that show they are gaining a sense of how words go together to create sentences (syntax) and how the same words can have different meanings depending on their order and other relationships (semantics). "Mama, breakfast" means something different from "Mama's breakfast," and so on.

Again, science has not yet solved the mystery of how this happens during a critical early developmental stage for toddlers. But we know that if either children's brains or their social relationships are insufficiently developed and nourished during the critical phase, language acquisition suffers. In the worst cases of either, including children growing up in abusive conditions of severe isolation from others, full language acquisition may not be possible. Extreme isolation from other people can be worse than a brain injury. The developing brain is so flexible in the way it organizes itself into networks of nerve cells that it is frequently possible for a child to gain full command of human language despite a severe brain injury — if the child is young enough and benefits from good language-based interactions with other people.

This picture suggests that dolphins and dogs, both highly social species, may make progress in language learning precisely because they can connect emotionally with people. The social character of these species and their cognitive evolution seem to be interdependent. Chaser's apparent edge on the dolphins in understanding sentences with three elements of grammar may result from the unique interspecies social relationship that has evolved between people and dogs. In any case I have no doubt that Chaser's focus on relating to people, a characteristic all domestic dogs share, is her strongest asset in learning language.

Support for that idea comes from language learning trials with nonhuman primates. The chimpanzee Washoe and the bonobo Kanzi, who were both raised and trained with lots of affectionate human contact, have far exceeded in language learning apes raised and trained without such contact.

Matching to sample is another way I am trying to use Chaser's emotional and social intelligence to advance her learning. When we do match-to-sample trials, I hide several objects — a newspaper, a washrag, a ball, a stuffed animal — in a corner of the living room, behind the couch, and in other places around the house. I'll show Chaser a duplicate of one of those objects and say, "Find this."

The only way I identify "this" is by pointing to it. Chaser has to recognize the most important characteristics of the object I'm holding and find another object with the same characteristics. At some level, conscious or unconscious, she must make the mental inference that I want an object like the one I am holding. In addition, Chaser has to understand that the object I am holding is not absolutely unique. She has to imagine that there can be others like it, and then hold that abstract idea in her mind while she searches for something similar. This may sound like a simple task, but rats and pigeons require hundreds of trials to learn to complete it.

Chaser's performance in matching to sample differs at home and at Wofford. At home, she has become very reliable in matching to sample, with results similar to those Juliane Kaminski has obtained in asking dogs to fetch an object after showing them a replica or a picture of it. But at Wofford, Chaser can be quite erratic in matching to sample. On some days, she quickly matches six objects in a row. On other days, she gets only three or four right. This is the one area of Chaser's training where there is such a difference with regard to place, and I haven't yet figured out what might explain it. In any case, match-to-sample learning remains a challenge as I try to identify what problem-solving hypotheses, implicit or explicit, might be in her head.

Imitation learning has also intrigued me for a long time, because to imitate something you have to be able to imagine yourself doing it. Both the imitation and the imagination can be unconscious. Young children generally don't realize they are imitating when they reflect behaviors they see their parents or other people display. Throughout life we consciously and unconsciously model ourselves on people who have a strong influence on us. But whether we realize it or not, we must make a mental inference about the point of view and intention of any individual we imitate before we can attempt to do as they do.

I mentioned that some scientists say animals cannot imitate because it requires the awareness that another individual has a unique point of view, supposedly something only humans have. This view of animals has always seemed wrong to me. Following Charles Darwin's example, I have no doubt that dogs are capable of empathy. Like imitation, empathy also requires imagination, an ability, conscious or unconscious, to recognize and share the feelings of another individual. I believe that as creatures capable of empathy, dogs must also have the imaginative capacity for imitation.

A few episodes with Chaser and other dogs heightened my interest in imitation learning. Since high school Deb has been best friends with Joyce Radeka, thanks in part to their mutual love of animals. Joyce and her husband, Frank Hodges, live in Columbia, South Carolina, and their house is full of their own pets and animals they are fostering for adoption. The summer when Chaser was a year old they offered to dog-sit her for a week while Sally and I stayed at their condominium in Myrtle Beach with Deb, Jay, and Aidan.

When we dropped off Chaser, we were warmly greeted by Joyce and Frank; Sky, a female Australian shepherd mix; Rudy, a large male Australian shepherd; and Don Juan, Don for short, a large male tuxedo cat. Don walked up to Chaser, who was still at my side, until their noses touched. After slowly sniffing Chaser's mouth, while she looked away,

Don brushed his body against her shoulder, and then turned his attention to the couch. He stretched his front paws up the back of the sofa,
as if for a fabric-ripping scratch.

Rudy caught this move, put his nose under Don's front legs to nudge
him off the couch, and then looked to Joyce, who responded with enthusiastic praise. Joyce explained that when Rudy joined their family
as a rescue dog, he observed her shooing Don away from the furniture. After seeing this only two or three times, Rudy delegated himself
to imitate Joyce's actions and keep the furniture safe from Don Juan's
mischief.

Before we left for Myrtle Beach, Joyce suggested a dip in their pool.
I seize any chance to get in the water, and it was clear that Rudy felt the
same way. As soon as he heard "swim," he was whining to get out of
the house and into the pool. I went to put on my swim suit, and when
I came out Rudy was in the water, paddling toward a ball. Chaser was
panting in the heat next to Sally, who was sitting under an umbrella
with Joyce.

Rudy walked up the steps at the shallow end of the pool, dropped
the ball at my feet, and shook himself vigorously, spraying water like
a garden sprinkler. I tossed the ball back in the water, and Rudy ran
around so that he could launch himself into a belly flop near the ball. I
joined him in the water, and after quite a few ball tosses and belly flops
I looked over at Chaser. Yasha loved the water, and I was hoping that
Chaser would follow suit. Although she was watching me carefully, she
showed no inclination to get wet. I called her over, but she wouldn't dip
a paw in the pool.

Getting back in the car, Sally and I felt a twinge of guilt and anxiety.
When we looked back, we saw Chaser staring after us with her ears
and tail down.

Fortunately we could not have asked for better dog-sitters. Joyce
e-mailed us pictures of Chaser with daily reports on their walks and
other activities. Still, when Sally and I walked in the door of Joyce and
Frank's house a week later, I thought Chaser was going to wriggle out
of her skin with joy.

Joyce invited us to stay for lunch and another dip in their pool. Over
lunch Joyce said that Rudy had belly-flopped into the pool every day,

that Sky had gone in the water via the steps, but that Chaser had only watched. Disappointed, I figured Chaser just wasn't a water dog.

When we went out to the pool, Rudy immediately belly-flopped in. After I got in the water, I turned to Chaser and said, "Chaser, are you going to come swimming?" With great caution, she walked down the steps at the shallow end, and then swam to me. She wasn't thrilled about it at first, but she did what she'd seen Rudy and Sky do every day of the past week.

Swim, that is, not belly-flop. Belly flopping was not Chaser's style, but it didn't take long for her to find joy in the water. The next time we went to the beach with Deb, Jay, and Aidan, we stayed at a pet-friendly motel and Chaser was in the surf with Aidan at every opportunity.

I also observed Chaser being a model for another dog. That same summer we took her to the Cherokee Mountains in North Carolina, where Robin was working her job as head rafting guide. It had been a while since Robin's dogs Blue and Timber passed away, and Robin had a new Siberian husky she named Spirit.

Robin had saved Spirit from being euthanized at the animal shelter, where she'd been abandoned by a puppy mill that no longer had use for her. Spirit was profoundly fearful of people. It was impossible for anyone but Robin to get close to her without her cowering and trembling in a way that broke your heart. Chaser and Spirit acknowledged each other but kept their distance, and we didn't push them to be friends.

When the rafting season ended, Robin came back to Spartanburg to spend some time with the family, and she and Spirit temporarily moved in with us. It was wonderful to see Spirit's new calmness. She seemed to be purring as she cuddled beside Robin. Sally and I tried to engage with Spirit, but she was still untrusting of anyone but Robin.

The depth of the abuse Spirit had suffered hit me on a fall afternoon when I was in the yard throwing a ball for Chaser. Spirit was watching from the porch. I called, "Spirit, catch the ball," and threw it to her. Spirit didn't budge as the ball bounced near her and hopped across the porch. In as encouraging a tone as possible I said, "Get the ball, Spirit!" She looked away.

The sadness I felt for Spirit and my appreciation for Robin's rehabilitation efforts both increased. I had never seen a dog completely ignore a bouncing ball and refuse all entreaties to play. The fact was that Spirit didn't know how to play. As far as Robin or we saw, she never even wagged her tail. I was afraid Spirit's inner light had been snuffed out permanently.

Robin continued to love on Spirit, however, and we kept cheering Spirit as best we could. We noticed Spirit watching Chaser play with her toys day after day, and Robin was sure this was a hopeful sign. One day while Robin, Sally, and I were all in the living room, Chaser was at the top of the stairs with a ball and Spirit was at the bottom. Chaser tossed the ball down the steps. Without moving, Spirit watched the ball bounce and roll to a stop.

Spirit timidly picked up the ball in her mouth and looked at Chaser. Not only that, but Spirit wagged her tail. We all had tears in our eyes at her first tentative attempt at playing. Since then, Spirit has become comfortable, if still a little clumsy, in imitating Chaser's play, and the two of them have become good friends.

Science is recognizing increasing numbers of examples of empathy and imitation among animals. For example, in their paper "The Evolution of Imitation," Ludwig Huber, Ádám Miklósi, and their colleagues have surveyed and analyzed imitation learning among birds, dogs, dolphins, fish, and primates. One of their major findings is that all animals tend to leave out details when they imitate a behavior. Like toddlers, they focus on what they perceive to be the goal of a sequence of actions and skip steps along the way. Related to this is that animals find it easier to imitate actions that involve an object, such as putting a ball in a tub, rather than pure body movements.

There is an immense amount to learn from the latest research. There is also a lot to learn about dogs' ability to mirror us through empathy and imitation from shepherds and others who work with dogs on a daily basis. In his book *Border Collies in America,* Arthur Allen wrote that owning and working a great herding dog "gives you a thrill that cannot be explained. I have worked dogs at trials that seemed to understand my every thought and movement. The elasticity between myself and dog was so complete that I thought my dog felt my every heart beat." Although Allen was not referring to imitation, I am sure that research on imitation learning can benefit from his emphasis on the emotional connection between dog and shepherd.

Wanting to capitalize on imitation for Chaser's learning, I began with behaviors she already knew on the basis of a verbal cue. For example, I'd say, "Chaser, do what I do. Watch Pop-Pop," and I'd sit down while saying "sit." Then I said, "Now you do it." Or I'd lie flat on my belly while saying "Lie down," and then add, "Now you do it." If she didn't sit or lie down, I repeated "sit" or "lie down." With many repetitions, Chaser began to grasp that "do what I do; watch Pop-Pop; now you do it" meant she needed to duplicate my actions, which might involve doing something with her toys, such as picking up a Frisbee and putting it in a tub, or might only involve body movements. Once she understood the "do what I do" cues, I could fade out other verbal or visual cues.

My initial goal was to reach a point at which I could introduce a behavior by saying no more than "Chaser, do what I do; watch Pop-Pop; now you do it," and providing only minimal verbal or visual cues after that. We were able to demonstrate significant progress in that at the APA in Washington, D.C., when Chaser imitated behaviors she was extremely unlikely to emit on her own, such as walking around the room in a precise pattern.

As Chaser's ability to learn by imitation improves, I won't always keep verbal cues, such as "Reverse," "Again," or "Faster," and visual cues, such as pointing, to a minimum. Instead, I envision combining imitation with verbal and visual cues for rapid learning, including the details, of complicated behaviors.

Progress depends on keeping Chaser interested in imitation learning. The most interesting imitation I've devised for Chaser, from her point of view, seems to be our stepping game. I move twenty or thirty feet away and we stand facing each other. I take one step forward and stop. Then I take another step forward with my other foot and stop. Chaser has to mirror me step for step. We move closer and closer, and then we move apart again with one backward step at a time.

Chaser likes this game a lot. But when we get about five feet from each other she no longer wants to step forward. She only wants to step backward. I could insist on closing the distance, but I respect her wishes. As time goes on, I hope I'll observe something that tells me more about this little idiosyncrasy of hers.

Because I want the imitation to stretch Chaser's creative imagination, I often demonstrate a behavior a few times in a row before I say, "Now you do it." By asking her to watch me a few times first, I'm trying to stimulate a mental rehearsal, which is in many respects equivalent to physical rehearsal. When we watch someone do something, or simply imagine ourselves doing it, our brain activity displays much the same pattern as if we were doing the behavior.

The challenge for me now is devising worthwhile behaviors for Chaser to imitate. I want to give her things to do that will enlarge her repertoire of useful behaviors and extend her imaginative capacity. The best one I've found so far may be a stroke of serendipity.

The Wofford head football coach Mike Ayers is an advanced black belt in the Tang Soo Do form of tae kwon do. A few years ago he let me take the class he teaches in it, and ever since then Chaser has loved to see me wave my arms above her head as if I am blocking and punching.

Even more exciting is when I go through the full movements. After a language lesson, match to sample, and imitation in our early-morning visits to Wofford, it's time for my workout. First I stretch in the big room where we've been working. While I'm lying down on a mat, Chaser keeps bringing a racquetball over. It's an occasion to reinforce her understanding of "closer," and when she drops the ball near enough I toss it across the room for her.

After I finish stretching I do my Tang Soo Do, striving to get less awkward at the movements. Chaser treats this as a special play period. Ball in her mouth, she darts this way and that, anticipating my steps, turns, and reversals. She mirrors my blocks, punches, and kicks as if she is my sparring partner. Maybe the better analogy is dancing partner. She whirls and bounces around me, lighter and surer on her feet than I can ever be on my mine.

At such moments it seems, to echo Arthur Allen, as if Chaser feels my every heartbeat. I never said, "Chaser, do what I do." And she is not doing exactly what I'm doing. But her mirroring of my actions goes beyond spontaneous responses. She knows what I am going to do from start to finish, and she has apparently built up a picture of the Tang Soo Do movements in her mind that is as clear and detailed as my own, if not more so.

Chaser's learned this all of her own volition, solely for the fun she can have with me. The unpredictable rewards we've experienced along the way have, I believe, made us both more resilient, persistent, and creative in pursuing our goals.

Aside from sharing life with Sally and me, one of Chaser's major goals is, no surprise, to turn every person she meets into a new friend and playmate. A week after her gratifying encounter with the women's soccer team, Chaser and I were finishing up a morning session at Wofford. We had just done our Tang Soo Do sparring dance, and I was using the exercise machines. When I moved from one machine to another, I kicked or threw her ball for her, but when I was on a machine she had to look for someone else to play with her.

I said earlier that no one's ever resisted Chaser's charm. I have to qualify that. No one's ever resisted her if she's had enough time to break through their emotional barriers.

On this particular morning four women students came in separately to exercise. As each one came in, Chaser approached her to initiate play. But each of them ignored her. She redoubled her efforts, maneuvering to put herself in the line of sight of one of them. She dropped her ball at their feet and wagged her tail and looked as cute as possible. No dice.

At one point she dropped her ball on the treadmill that one young woman was using. I said, "Chaser, no." That was not something I wanted her to do again. But she was unhappy with the result anyway, because the ball bounced off the moving treadmill and she had to find it between a couple of other machines. If I set up a tennis ball machine for Chaser, she would not be happy chasing the balls it shot out. She wants to be doing something with and for a person.

For more than thirty minutes, Chaser went from one young woman to another with her ball. None of the four ever stopped to interact with her, but she didn't stop trying.

Finished with my workout, I called to Chaser, "Come on, girl. Time to go." But she was focused on one of the young women she wanted to win over. She would come in a flash to "Here!" But I like what Chaser and I do together to be a shared idea whenever possible. So instead I said, "Chaser, let's go see Nanny."

The thought of seeing Sally immediately made the hard-to-get girls in the exercise center uninteresting, and Chaser came to my side with her tail wagging. The day has not begun properly until we get home from Wofford around seven a.m. and Chaser goes into the bedroom to see Sally, who is usually just getting up. As soon as Sally makes the bed and tells Chaser it is okay for her to get on it, they play catch with one of her toys. After that we all have breakfast together, and then Chaser takes a little nap before our next language training and play session.

The next morning at Wofford I was on a treadmill when a different young woman came in and started using the treadmill three stations away. Chaser went over, and instead of dropping her ball on the floor or on the treadmill, she simply held it up in her mouth, lifting her head as high as she could. I glanced away, and the next thing I knew the ball was flying across the room with Chaser in happy pursuit. She brought the ball back to the young woman on the treadmill and again held it up in her mouth. This time I saw the young woman look down at Chaser, smile, and then take the ball from her mouth and throw it for her again.

I love seeing Chaser's determined spirit. She shows it every day in her willingness to keep at the tasks I set for her, as well as in pursuing her own goals. One of our training games at Wofford involves a large exercise ball — Big Ball, we call it. We work with Big Ball in the room between the racquetball courts and in the hallway on the other side of the room from the main exercise area.

"To Pop-Pop, take Big Ball," I tell Chaser. She primarily understands "take" to mean "pick it up in your mouth," but this inflatable exercise ball is twice as big as she is. She can't hold the ball in her mouth, and the only way she can really move it effectively is by nosing it.

"To Pop-Pop, take Big Ball" is a lesson in how the meanings of words can be ambiguous and depend on context. It's also a lesson in persistence.

For some time Chaser's response to "To Pop-Pop, take Big Ball" was to nose the ball in almost any direction. As she did this I repeated, "To Pop-Pop, take Big Ball," until the ball finally wound up near me, often as much by accident as by her efforts.

But when I said, "To Pop-Pop, take Big Ball," one day in early 2013, Chaser trotted over to the ball, some thirty feet away down the long hall, and paused. She cocked her head, looked at the ball, looked back at me, and then went behind the ball and nosed it deftly straight to me.

We celebrated with play. And then I set the ball farther away and said, "To Pop-Pop, take Big Ball." Again she went behind the ball and nosed it straight to me.

We celebrated again with play. I set the ball even farther away. And for the third time she responded unerringly.

The next day she was again perfect. She had not only learned a physical behavior. She had consolidated a concept in her mind, and that made a previously difficult behavior much easier for her.

There is still a little bit of a delay when I say, "To Pop-Pop, take Big Ball." She has to work through what I mean by "take" before she starts trying to nose it to me. And there's another difficulty too. Big Ball can easily get lodged in a corner or behind another object. When this happens I see and hear her frustration. She starts barking impatiently, as if to say, "This is impossible! Help me out here!"

All I say is, "To Pop-Pop, to Pop-Pop."

It's very frustrating for her, but she keeps nosing and pawing at the ball. Eventually she finds that the task isn't impossible. She gets the ball rolling again and noses it toward me. But it often gets hung up again, depending on how the exercise mats and other things are situated. Again she barks loud and clear, "Help me out! I can't do it! This is impossible!"

Again I say calmly, "To Pop-Pop, to Pop-Pop."

She continues to bark that she needs my help, but she never quits on the job. Sooner or later she noses and paws Big Ball out of the last spot where it's gotten lodged, and she brings it all the way to me. "You did it! Good girl, good dog!" I tell her, and we celebrate her triumph as always with play.

And so, game by game, day by day, we continue to expand our joyful conversation.

Epilogue
Unleashing Chaser's Genius

TWO-YEAR-OLD JASON, grandson of our next-door neighbor — and Sally's fellow Ya-Ya — Theresa, throws a pinecone in our front yard. His throwing motion is not as fluid as it soon will be, but it's a good heave for a toddler, and the pinecone lands on the grass about four feet away from him.

Nine-year-old Chaser watches Jason out of the corner of her eye as she begins to creep toward the pinecone, moving as slowly as when we play our mirror-stepping game.

Jason grins and runs for the pinecone, his chubby arms swinging his torso back and forth, his little legs still getting used to balancing him at that speed.

Chaser lunges forward at the last second, snatches up the pinecone in her mouth, and races away, with Jason in gleeful pursuit. But he can't catch up with her and he soon stops running, a hint of frustration coming into his expression. Seeing this, Chaser stops, half turned toward Jason. And then she quick-steps over to him, drops the pinecone at his feet, and retreats a little, waiting for his next move. A grin spreads back across Jason's face as he bends at the middle — so sharply, that for a fraction of a second I fear he will tumble over — to pick up the pinecone. He can just get one hand around it, and raising it high he runs to the side of the yard, while Chaser follows at a bit of an angle, as if arcing out around some sheep.

Jason winds up and throws the pinecone as far as he can, a good foot farther than before, and he and Chaser continue their game. They will play like this, never tiring of the repetition of theme and variations, until the proverbial cows come home or Theresa says, "Come on, Jason, it's time for your nap."

Although Chaser loves children, she keeps her distance from the littlest ones, partly as Sally and I taught her when she was a young puppy, so that she never unintentionally disturbed a baby or a protective parent. But the behavior also comes from knowledge, gained by experience, that very young children tend to yank and pull on her and can't run or throw very well.

That was how Chaser behaved around Jason when he was always in his stroller outside or just beginning to walk. Now that he is getting his legs under him, the dynamic has changed, and I've been watching them learn from each other. Jason has discovered that pulling Chaser's coat, ears, or tail makes her move far away and not want to come back. Together they have figured out how to adapt games such as catch and keep-away to Jason's fast-advancing mental and physical skills.

Jason is in awe of Chaser. He loves to see her leap for the Frisbee and catch it in midair. For her part, she gets more interested in him as a playmate every day, as his increasing coordination and motor skills cue up his cognitive development. There is quite a friendship in the making here, and at eighty-five I am encouraged to think that in the future I might offload some of Chaser's exercise demands onto my young next-door neighbor.

For the present they are two of a kind, the human toddler and the dog that will always be a toddler emotionally and cognitively, the eternal puppy. Flexible in body and spirit, they each exemplify and express a 360-degree gradient of curiosity about the world and openness to new experiences.

In contrast to their wolf ancestors and to wild dogs, genetic selection via domestication has given domestic dogs an enduring juvenile persona. Their always-a-puppy-at-heart nature evokes our human need to nurture, and it also keeps dogs in a state of readiness to learn new things. Harsh experience can bury the puppy self deep inside a dog, but love and play can almost always reawaken the puppy within, as Robin's love and Chaser's example of play gradually reawakened the playful puppy inside Spirit.

I believe that it is my daily play with Chaser that sustains her inner puppy as the source of her creative energy and ability to learn. Likewise, I believe that it is Chaser's puppyish play with me that sustains my inner child as the source of my creative energy and ability to discover things with her. During the years when Sally and I did not have a dog, I felt my playful side diminish. With Chaser as a member of our family, however, I once again "carry mischief in my pocket," as Sally, Robin, and Debbie like to say.

What I am asserting here about play and creative learning potential may not yet be a testable scientific hypothesis, but I am sure that eventually it will be. And I believe with every fiber of my being that testing will confirm and extend the hypothesis that the most profound learning is impossible without play. In chapters 7 and 16, I discussed an essay on creative learning by the great animal scientist John Staddon. During his doctoral studies, Staddon was one of B. F. Skinner's teaching assistants. But the title of the essay, "Did Skinner Miss the Point About Teaching?," indicates how Staddon has charted his own path in science. He took the best of Skinner's paradigm and reinvigorated it based on his own and other scholars' research into animal learning.

In his essay, Staddon illuminates Skinner's great insight that in operant conditioning, which we might better call operant learning, the learner's voluntary, unconstrained responses are "essentially spontaneous, at least on first occurrence" and that reinforcing these responses makes them richer and more varied. This relationship between reinforcement and freely chosen operant responses is ultimately what makes creative learning creative.

Skinner's idea of reinforcement and unpredictable, free-operant responses is akin to Darwin's idea of selection and variation in the evolution of species. In his own research and teaching, Skinner focused on reinforcement, and so has much animal science ever since. But Staddon calls on animal science to switch focus and "look around for the sources of variation that yield the most exciting kinds of teaching [and learning]." The most powerful such source of variation, the best nurturing for creative learning, is surely play.

As I also discussed, the most powerful paradigm in animal science has come down to us from Descartes, the seventeenth-century philosopher who decreed, without any experimental evidence, that animals are flesh-and-blood machines that cannot think or feel—and thus obviously cannot play, either. Skinner moved away from this paradigm with his conception of free-operant responses to new situations, and Staddon among others has demonstrated that virtually all animal learning involves conscious or unconscious mental inferences. But Descartes's animals-are-meat-machines paradigm remains extraordinarily powerful.

A paradigm in science holds sway so long as the scientific consensus can ignore or dismiss anomalies that contradict it. But the fate of every paradigm in science is sooner or later to be abandoned, or significantly revised, in the light of persistent anomalies. With her language learning, Chaser joins the chimpanzee Washoe, the dolphins Phoenix and Akeakamai, the African gray parrot Alex, the bonobo Kanzi, and her fellow Border collie Rico as anomalies that can no longer be dismissed or ignored. Together they tell us that it is past time to abandon Descartes's paradigm of animals as machines and to replace it with a new paradigm of animals as truly our fellow creatures—biologically, emotionally, and cognitively.

Capitalizing on Chaser's super-social nature and her talent for listening, herding, and creative play has opened up her ability to understand language. Together, she and I have found her light and let it shine.

I hope that Chaser's story will inspire play-based training of dogs and other animals by scientists and nonscientists alike. If such training becomes commonplace, it can unleash the genius of all dogs and many other species, and in so doing expand humanity's horizons to encompass a new understanding of ourselves in the natural world.

In the near term, Sally and I are looking forward to relaxing on a sunny beach. Of course Chaser will accompany us, so that she can run on the beach, play with her toys, especially the balls and Frisbees, and swim in the ocean. That's the plan, anyway. But as I've found again and again in life, including writing this book, you never know what the next chapter will be.

Chaser is in her Border collie prime. If she can't be herding sheep, she needs to be gathering up her toys and solving problems in connection with them. That's what keeps her mind and body engaged. That's what keeps her learning without knowing she's learning. That's what keeps us both progressing along the road to understanding the mysteries of learning for both humans and animals. Our journey of discovery is nowhere near its end, and we're counting on lots of fun ahead.

Here's Chaser coming up to nudge me as I sit at my desk. She drops a ball at my feet and locks her eyes on mine.

"One, two, three!" I throw the ball, and the game is on.

Afterword

A Future of Many Chasers

IN THE YEAR since this book was published in hardcover, I've been excited to see how Chaser's learning fits with other recent findings in the field of animal cognition. For example, on November 18, 2013, the *New York Times* reported on research, first published in the scientific literature a few years earlier, showing the learning abilities of tortoises and lizards. The article's headline, reflecting a reversal in the prior scientific consensus on these creatures, was "Coldblooded Does Not Mean Stupid."

I can't resist quoting the article's opening words: "Humans have no exclusive claim on intelligence. Across the animal kingdom, all sorts of creatures have performed impressive intellectual feats. A bonobo named Kanzi uses an array of symbols to communicate with humans. Chaser the border collie knows the English words for more than 1,000 objects."

It was fun to see Chaser cited as a super achiever along with Kanzi. It was also fascinating to read that a team at the University of Lincoln in England has documented the ability of tortoises to navigate a maze to find hidden strawberries. As Anna Wilkinson, the leader of the research, told the *New York Times,* doing this efficiently "requires quite a memory load because you have to remember where you've been."

The comment struck a chord with me because Chaser has certainly needed to draw on an extensive memory system in her learning.

The article also reported on a separate finding by Wilkinson and her research associates, that tortoises can learn by imitation a task that they seem unable to learn on their own. When the study's tortoises needed to navigate around a fence to reach some food, none learned to do it spontaneously on his or her own. But all the tortoises in the study learned to do it by watching a trained tortoise complete the task.

As I discuss in chapter fifteen, the ability to learn by imitation seems to rest on a theory of mind, the concept that there are other beings in the world with their own points of view. Young children display an implicit theory of mind, without conscious awareness of the concept, when they begin to understand what other people are communicating through words and through gestures such as pointing. The tortoises' ability to learn by watching a trained tortoise also displays an implicit theory of mind. That's what Chaser displays, too, when she imitates my actions in response to my verbal and visual cues.

In February 2014 *Scientific American* reported on a wealth of recent research on elephants. The traditional folk wisdom that elephants never forget turns out to be well rooted in the facts: the creatures really do have remarkable memories. Also admirable and worthy of further research are their cooperative social behaviors and problem-solving skills, their subtle communication through a wide range of sounds and body language, their adaptive tool use, their apparent self-awareness, and their profound empathy for one another.

One of Aesop's fables, "The Crow and the Pitcher," tells of a thirsty crow who figures out he can reach the water in a nearly empty pitcher by dropping pebbles into it to make the water rise. Experiments at the University of Auckland in New Zealand have found that at least some real-life crows are just as smart. Presented with the challenge of getting food floating out of reach in a container of water, New Caledonian crows adopted the same strategy as the one in the ancient fable: they dropped pebbles into the container to raise the floating food high enough for them to pick it up in their beaks. In a March 2014 article in the online scientific journal *PLOS One,* the researchers reported that New Caledonian crows can solve this problem about as well as five-to-seven-year-old children can. To do this, the crows must draw implicit inferences about cause and effect. Their ability to do so is close kin to Chaser's ability to draw an implicit inference in learning an unfamiliar toy's name by exclusion.

The learning research going on with crows these days is highly varied. A website devoted to it is called cooperativecrows.com, a name that shows animal scientists are finding the social intelligence of crows as remarkable as the social intelligence of dogs.

I'll also mention a continuing line of child learning research, which I first discuss in chapter four, showing that there is a huge gap in cognitive development between young children whose parents speak to them frequently in positive ways on a variety of subjects and those whose parents do not. Just as this book was being published in hardcover, the *New York Times* reported on research at Stanford that found a six-month lag in language skill development between two-year-old children in these two different groups. The thrust of this and similar studies is that talking a lot to young children effectively tunes up and supercharges their cognitive development by recruiting their innate proclivity for social intelligence.

The positive news for all children in this and similar findings is that parents can learn to talk to their young children in a way that facilitates their acquisition of language skills and their general cognitive development. This reminds me of how I formed the hunch that talking to the puppy Chaser throughout the day just as I would speak to a toddler might help her learn basic elements of language. I can't be sure that all my talking to Chaser as a puppy was beneficial to her subsequent language learning. But I strongly suspect it was.

With regard to child learning, I was also intrigued by a study showing increased activity in specific brain areas when children learn by imitation. Researchers at the University of Washington announced in the fall of 2013 that they had succeeded in mapping these areas by having toddlers wear caps with electronic sensors. It's a job for a researcher more technologically adept than I am to fashion such caps for dogs, but it would be fascinating to see if, as I suspect, there is increased activity during imitation learning in the corresponding areas of dogs' brains.

There are striking common threads in these studies and my experience with Chaser. What leaps out most strongly, in my view, is the power of social learning and of the implicit conceptual understanding that social learning alone may support. It is fascinating to see social learning occurring even in "nonsocial" animals like tortoises and lizards, with their supposedly primitive "reptilian brains."

Two heads really are better than one, especially when they're joined in empathy and play. Many dog lovers have long felt that their dogs yawn in sympathy with them. A recent University of Tokyo study, reported in the media while the hardcover edition of this book was going to press in August 2013, demonstrated that dogs do indeed "catch" their owners' yawns and yawn in response. The researchers also found that dogs preferentially catch the yawns of their owners over those of strangers, which is super strong evidence of dogs' capacity for empathy.

All these scientific findings reinforce the belief I share early on in the book that sooner or later there will be many Chasers, including both dogs and members of other species that humans bring into their lives as companions, pets, and working partners. The media's continuing interest in Chaser and readers' enthusiastic reactions to the book also encourage me to believe this.

When the book was published in hardcover, it got a really awesome response from bloggers, the media, and readers. One memorable television appearance came when I was under the weather and couldn't go on *Fox and Friends* with Chaser. My younger daughter, Debbie, stepped into the void and took Chaser on the show. I was so proud of them both as they demonstrated Chaser's language training, especially her ability to learn by exclusion with a toy that host Steve Doocy supplied. Articles and blog posts in a wide range of publications and websites, including *Dog and Hound, People, Popular Science, Scientific American,* the *New York Times* Sunday Style section, the *Huffington Post,* Time.com, and WSJ.com, reflected the curiosity that people around the world have about Chaser and her learning.

Many readers have e-mailed me about the book, but I'd like to share a note that was sent not to me, but to Chaser, through her Facebook page:

Dear Chaser,

Just a little note to let you know how much I am enjoying reading about you . . .

I love the way your Pop-Pop writes about his relationship with you . . . and reading about your work [with Dr. Pilley] helps me train [our new dog] . . .

Dr. Pilley's observations about his research with you confirm what I have learned with [my dog] and augment the joy I feel in living and working with this wonderful companion. Thank you for inspiring so many people and dogs . . .

This is the first fan mail I have ever written, and I am fifty years old and you are a dog.

Yours very truly,
Kathryn M.

It is a great feeling to see from responses like this that the book has clearly communicated Chaser's story and the principles I followed with her. Occasionally, however, I hear from people who wonder if I have an extra tip up my sleeve I could share with them to use with their dogs.

I have put more or less all I know about dogs' learning into telling you about Chaser and showing you how I have interacted with her. The key ideas are all in the anecdotes I share about living and working with Chaser from the day she entered the Pilley family. Use play to motivate your dog's learning, and capitalize on the energy of play every step of the way you take together. If you want to teach your dog the basic elements of language, begin with play with one named object at a time. Gradually build up your dog's vocabulary of different types of words. And then work on combining those words into sentences with multiple elements of grammar, like "to Ball, take Frisbee" and "to Frisbee, take Ball."

If there's some other creative learning you want your dog or another animal to acquire, try to do the same things in those areas. Identify the ABC concepts or behaviors the learning requires. Find a way to give those things implicit value in play. And leverage the energy and inherent reinforcement of play to motivate solving a series of increasingly difficult challenges.

Fundamentally, as I hope this book has shown, there is one thing you must do to explore learning through play with a dog or other animal: respect the other mind involved with you in the process. Remember that communication is a two-way street, and allow for that other mind to influence you and your behavior. I experience this with Chaser every day as we continue trying to extend her learning through play.

My conscious focus in our various games is on increasing Chaser's ability to understand sentences with multiple elements of grammar and to learn by imitation combined with verbal and visual cues. However, the mutual conditioning between Chaser and me has become such that it's often not clear to me who is the teacher and who is the student in determining what games we play and for how long.

It's much easier for me to track who is leading whom when someone else is interacting with Chaser. Early in 2014 Anderson Cooper, producers Sumi Aggarwal and Denise Cetta, and a *60 Minutes* crew came to visit, and I had to laugh at a conversation between Denise and my older daughter, Robin. We had all just been working together inside for a while, with Chaser cooperating nicely in demonstrating her learning. Robin took Denise to the backyard to see Chaser's agility course, and Chaser followed with a ball in her mouth. Denise repeatedly asked Chaser to bring the ball to her, but Chaser repeatedly ignored her.

Denise asked Robin, "What's going on? Am I not saying it right?"

Robin said, "Chaser's not necessarily the most obedient dog. She's an independent thinker. She's been doing things our way all morning, and now she wants to do things her way. Right now, she probably wants to play keep-away and have you chase her."

Even though she is a mature dog of ten, with a touch of arthritis in one hip, Chaser remains a puppy at heart, an eternal toddler. And like toddlers, her favorite word is "No!" She can't verbalize it, but she can express it through her behavior. It's one of many ways in which Chaser shows that dogs think and feel much as humans do.

Of course, Chaser remembers her obedience lessons from when she was a puppy and we were first teaching her not to run after cars or go into the road for any reason without permission. She respects the gravity of a serious tone of voice, and she immediately responds to an obedience signal with the appropriate behavior in these situations. I tease Robin and Debbie by saying that Chaser's responses at such times put her way ahead of them when they were willful toddlers. Robin teases me right back by saying the real difference is that I've gotten better at working with toddlers.

Chaser's learning and the other new findings on animal intelligence point to the fact that we can't treat ourselves right if we treat animals wrong. It makes me very happy that Chaser has become a landmark personality in our growing understanding of what animals share with humans as thinking, feeling beings. In this connection I love the words of Martin Luther King Jr., who said, "Never, never be afraid to do what's right, especially if the well-being of a person or animal is at stake. Society's punishments are small compared to the wounds we inflict on our soul when we look the other way."

I am glad to see the growth of animal studies as a new academic discipline devoted to understanding the animal-human relationship. Work in animal studies straddles the border between science and the humanities and will, I believe, help establish more common ground between them. I have long believed that fully understanding anything fundamental, in human life or in nature as a whole, requires us to combine three ways of experiencing and knowing the world: science, myth, and poetry. If we don't have all three, we're bound to miss something important, and it's in this spirit that I've sought to tell Chaser's story.

Sooner or later, there will be many Chasers. Now that Chaser and I are both in our later years, we're impatient to see other animals follow in her footsteps and perhaps go beyond her. So I hope you'll seize opportunities for learning through play in your own relationships with dogs and other animals. Who knows? You may be living with a puppy or other creature who can, like Chaser, extend the boundaries of what we thought was possible for animals and expand our understanding of nature and our rightful place in the natural world.

Please let Chaser and me know about your experiences by writing to us through our website, chaserthebordercollie.com. In the meantime, here is a big hurray from me, and a tail wag and a woof from Chaser, for everything that dogs and other animals will help us discover about learning in the years ahead.

―

Chaser's Toys

This list includes toys whose names Chaser has learned in addition to the 1,022 toys that Alliston Reid and I discussed in our 2011 paper in *Behavioural Processes*. It also uses my frequently phonetic spellings of names as I have written them on Chaser's toys.

A: ABC, Acorn, Afro, Ahab, Aidan, Al Jolson, Alleycat, Alligator, American, Amphibian, Andy, Angel, Ann, Ant, Antler, Ape, Apple, Apple Ball, Applejuice, Arf Arf, Armadillo, Asp

B: Baa Baa, Babe Ruth, Baboon, Baby, Baby Bear, Baby Clown, Baby Dolphin, Baby Elephant, Baby Moose, Baby Octopus, Baby Panda, Baby Polar Bear, Baby Shoe, Baby Sleeping, Back Door, Back Porch, Bacon, Badger, Bagel, Baker, Balloon, Bambi, Bamboozle, Bar, Barbell, Barker, Barney, Barrell, Baseball, Baseball Player, Basket, Basketball, Bass, Basset Hound, Bat, Batman, Baton, Beach Ball, Beagle, Bean Ball, Bear, Bearing, Beautiful Dreamer, Beauty, Bed, Beep Beep, Bell, Belt, Belzubub, Bench, Benji, Bessie, Bessie 2, Bessie Bee, Bicycle, Big, Big Ant, Big Coke, Big Cotton, Big Elephant, Big Football, Big Frisbee, Big Gator, Big Golfball, Big Holey Ball, Big Pillow, Big Rope, Big T, Bike, Billy Goat, Bimbo, Birdhouse, Black Bear, Blackie, Blanket, Blind, Block, Blondie, Blue Ball, Blue Bird,

Blue Frisbee, Boar, Boat, Bob, Bobbing Cork, Bobo, Bomb, Bone, Boney Ball, Bong, Bongo, Boo, Boob, Boomerang, Boot Ball, Boss, Boston, Boston Ball, Bouncy, Bovine, Bow, Bowlegs, Boxer, Boy, Bracelet, Brer Rabbit, Briarpatch, Broken Bone, Brown, Bruin, Brush, Buck, Bucket, Buddha, Buddy, Buffalo, Buffoon, Bugaboo, Bugbox, Bugs Bunny, Bullseye, Bullwinkle, Bumble Ball, Bumble Bee, Bump Ball, Bunny, Bunny 2, Buoy, Burrow, Buster, Busy Bee, Butch, Butterfinger, Butterfly

C: Cactus, Calf, Calico, Camel, Candy, Cantelope, Cap, Captain Hook, Car, Carp, Carrot, Casanova, Case, Casper, Cat, Cat Ball, Catcher, Caterpillar, Cellphone, Centipede, Cha Cha, Chair, Charlie Brown, Checkers Ball, Cheerleader, Cheese, Cheeseburger, Cheetah Frisbee, Cherry, Chewey, Chewstick, Chi, Chicken, Chicken Little, Chimp, Chimpanzee, China Ball, Chipmunk, Chippy, Choo Choo, Chubby, Chuckit, Chuckles, Cinderella, Circle, Circus Elephant, Cleanipals, Clock, Clodhopper, Cloth, Clothespin, Clown, Clownfish, Club, Cobra, Cobra 2, Cockadoodle Doo, Coffee Table, Coke, Collar, College Ball, Colt, Comeforth, Connor, Cookie, Coon, Coop Ball, Coral Snake, Core, Cork, Corn, Cottontail, Couch, Cow, Cow's Hide, Cow's Hoof, Crane, Crawdad, Crawfish, Crawler, Creepy, Croc, Crocodile, Croquet Ball, Crow, Crow K, Crushed Ball, Cry Ball, Crybaby, Crystal Ball, Cub, Cube, Cuddler, Cupid, Curly, Curtain, Cushion

D: Daddy Longlegs, Daffodil, Daffy, Dairy, Daisy, Dakota, Dapper Duck, Dart, Darwin, Debbie, Decoy, Deer, Demon, Die, Dimples Ball, Dinasaur, Ding Dong, Dinosaur, Doe, Dog, Doggie News, Dolphin, Dominica, Donald, Donkey, Donut, Doddlebug, Dopey, Dora, Dory, Double O, Downstairs, Dozer, Dragnet, Dragon, Drake, Dreamer, Dresser Upper, Drumstick, Duckie, Duckling, Dumbo, Dutchboy, Duvet

E: E, Ears, Easter Bunny, Easter Egg, Easter Egg Ball, Eel, Eightball, Elephant, Elk, Elmo, Elsa, Ernie the Worm, Eros, Escargots, Ewe, Exotic Bird, Eye

F: Fan, Farmer, Fatty, Feet, Feline, Felix, Fido, Figment, Fire Hydrant, Fireball, Fireman, Flagball, Flatfoot, Flex, Flintstone, Flip Flop, Flipflopper, Flippity Flop, Flo, Float, Floozy, Floppy Ear, Flyfish, Foam Ball, Foam Rope, Foodbowl, Fool, Fool's Gold, Foosball, Foot, Football, Forty-Seven Ball, Fox, Frame, Franklin, Frat Rat, Free Willy, French Fries, Fritz, Front Door, Front Porch, Frontline Frisbee, Frosty, Furby, Fuzzy, Fuzzy Frisbee

G: Game Ball, Gander, Garfield, Gary Giraffe, Genie, Gensing, Geoffrey, Georgia, Ghosty, Ghoul, Giggly Wiggly, Gingerbread, Ginny, Giraffe, Girley, Glass, Globe Ball, Glove, Goat, Godzilla, Gold Ball, Goldfish, Golf, Golf Ball, Goo, Goofy, Goose, Gorilla, Gosling, Gourd, Grass, Grasshopper, Green Ball, Green Snail, Groovy, Groucho, Growl, Grumpy, Gum Gum, Gus

H: Hair, Half Blue, Halloween, Hamburger, Hammer, Hand, Hand 2, Hankerchief, Happy, Hard Ball, Hare, Harness, Hassock, Hat, Headless, Health, Health Ball, Heart, Heavy Ball, Helen Keller, Helium, Helmet, Henry, Hobby Horse, Hobie, Hog, Ho-Ho-Ho, Holey Ball, Holiday, Holly, Holy Grail, Honeybee, Hoot Owl, Hoppy, Horn, Horse, Horseshoe, Hose, Hot, Hot Dog, Hot Stuff, Hot Tamale, Houdini, Hound, Hubby, Huey, Huggable, Huggers, Hugs and Kisses, Hula Girl, Hula Hoop, Hummingbird, Humpty Dumpty, Husky, Hut Ball, Hypo

I: Igloo, Iguana, I-Love-You, Indian, Infant, Inky, Insect, I-Wanna

J: Jack Ball, Jackass, Jackrabbit, Jaguar, Jaguar Ball, Jelly Roll Ball, Jersey Cow, Jester, Jingle Bell, Joe Cool, Joker, Jonah, Joy, Joy-Joy, Jug, Jumbo, Jumpy, Jungleboy, Junior

K: KG, Kangaroo, Kayak, Keg, Kermit, Key-Key, Kick Ball, King of Jungle, King Snake, Kitty, Kitty Kitty, Klutz Ball, Knife, Knob, Koala Bear, Kong

L: Ladybug, Lamaze, Lamb, Land Mine, Laughing Bear, Laughter,

Leaf, Leash, Leather, Leg, Lemon Ball, Leo, Leopard, Leopard Ball,
Lifesaver, Lightning Frisbee, Like-My-Hat, Lion, Lion King, Lips,
Lipsey, Lipsmacker, Liquid, Listerine, Little, Little Bouncy, Little Boy
Blue, Little Cottonball, Little Elephant, Little Football, Little Goofy,
Little Heart, Little Hula Hoop, Little Shark, Lizard, Lobster, Lofanzo,
Log, Loggerhead, Logs, Long Rope, Loop, Loudmouth, Love Ball,
Lover, Lucky, Lucy

M: M&M's, Macaw, Madagascar Ball, Magoo, Maid, Mailbox,
Mallard Duck, Mama Bear, Mammy, Man-on-Moon, Marble Ball,
Marvel Ball, Mary, Master, Mates, Meow, Mephistopheles, Merlin,
Mermaid, Michael the Angel, Mickey, Mickey Mouse, Mighty Mite,
Milk Ball, Minnie, Minnie Mouse, Minnow, Mischievous, Miss
Moose, Miss Mutton, Miss Piggy, Mister Potato, Mitten, Mmm,
Mo, Mo-Mo, Moby Dick, Model T, Mole, Molly, Mom, Mongrel,
Monster Ball, Monty, Moo, Moo Cow, Moo-Moo, Moo-Moo-Moo,
Moongirl, Moose, Mopsy, Mosquito, Moth, Mother Goose, Mother
Rabbit, Mountain Goat, Mountie, Mouse, Mouthy, Mule, Muppet,
Muscleman, Music Bear, Mutt

N: Nemo, Neopets, Nerf, Never Forgets, Night Ball, Ninny Ball,
Nipple Ball, Noisy, Noose, Nosey, Nun

O: Octopus, Odie, Oink, Old Navy, Orangatang, Orange Ball, Orb,
Osafa, Oval

P: Pachaderm, Pad, Pajama Girl, Pajamas, Pancake, Panda, Panty
Ball, Paper, Paper Bag, Parachute, Parrot, Party Girl, Pat the Bunny,
Patch Ball, Paws, Peachy Ball, Peanut Butter, Peanuts, Peep, Pegasus,
Peggy, Pen, Penguin, Peppermint, Pepsi Ball, Perch, Percy, Peter,
Peter Rabbit, Phone, Pica, Pickup Truck, Piggy, Piglet, Pigskin,
Pigtail, Pillow, Pinky, Pinnochio, Pipe, Pistol, Pixie, Planet, Plate,
Platypus, Playmate, Playpen, Plum Ball, Pokemon, Polar Bear, Pole,
Polka Dot, Polka Dot 2, Polly, Polo, Poltergeist, Pona, Pony, Poof Ball,
Poogie, Pooh Bear, Pooh Pooh, Popcorn, Popeye, Poppy, Porcupine,

Porkchop, Porky, Porpooise, Post Ball, Potholder, Powder Puff, Power Ball, Prancer, Precious, Pregnant, Pretty Cat, Pretty Girl, Primate, Prisoner, Prissy, Professor, Propeller, Pu Ball, Pucker, Puddle, Puddy Cat, Puddin, Puff Ball, Puffy, Pug Nose, Pug Nose Ball, Pumpkin, Punching Bag, Punt, Punt Ball, Pup, Puppy, Puppy Love, Purple, Purple Holey Ball, Purse, Pussy Cat, Pyramid

Q: Quack Quack

R: Rabbit, Raccoon, Race, Racket, Radar, Raggedy Ann, Rainbow Ball, Rainbow Fish, Raincoat, Ram, Rangaboom, Rattler, Rawhide, Red Ball, Red Dog, Red Riding Hood, Red Rope, Redneck Pheasant, Reebot, Reindeer, Reindeer 2, Rex, Rhino, Rhinoceros, Ribbon Ball, Ringer, Rip Van Winkle, Ripper Frisbee, Road, Roadrunner, Roadster, Roast Pig, Robin, Rocket, Rodent, Rodney, Rolls Ball, Rooster, Rope, RopeBall, Rough, Rough Dog, Royal, Rubber, Rubber Band, Rug, Rugby, Rusty

S: Sabu, Sadie, Sahara, Sammy, Santa Claus (a.k.a. Santie Claus), Santa Fe, Sassy Ball, Satellite, Scarecrow, Scooby Doo, Scotty, Sea Urchin, Seacow, Seal, Serpent, Shamu, Shark, Sharpei, Sheep Dog, Sheriff, Shnauzer, Shoe String, Shoes, Shorts, Shovel, Shrek, Siamese, Sidewalk, Sieve, Silent Night, Silver, Simba Ball, Simian, Ski Cap, Skinny, Skunk, Sky Ball, Slammer, Sleepy, Sling Ball, Sling Shot, Slipper, Slow Poke, Slug, Small, Smiley Ball, Smokey, Snail, Snakey, Snapper, Snicker Bar, Snoopy, Snow, Snowball, Soap, Soccer Ball, Sock Ball, Socks, Soft, Soft Balloon, Soft Blue, Softball, Softy, Sol, Sonic Ball, Spangle Ball, Spider Monkey, Spiderman, Spie, Spin Ball, Spirit Dog, Splash, Splat, Splat Ball, SpongeBob, Spook, Spoon, Sport Ball, Spot, Spring, Square, Squash, Squeeky, Squid, Squirrel, Squirrel Monkey, Squish Ball, St. Anthony, St. Nick, Stack, Staff, Stag, Stakes, Star, Steps, Stewart Little, Stick, Stick Ball, Stiltz, Story Book, Street, Stripes, Stubborn, Sugar, Sun, Sun Ball, Sunshine, Super Dart, Surf, Surf Ball, Sweatshirt, Sweet Potato, Sweetheart, Sweetie Pie, Swing, Swing Ball, Swing Zocker, Swish, Sylvester

T: Tadpole, Tangerine, Tank, T-Bone, Team Ball, Tee Shirt, Teen Titan, Teeth Bone, Telephone Pole, Teletubby, Tennis, Tentacle, Terrier, Thimble, Thomas, Tick Tock, Tie Face, Tiger, Timber, Tin, Tin Man, Tinkerbell, Tiny, Toad, Toby, Tomato Ball, Tomcat, Tongue, Tonka, Tony the Tiger, Toothy, Top, Top Cat, Top-It Ball, Tortoise, Towel, Tower, Trapeze, Treadwheel, Treasure Box, Treat, Tree, Triangle, Trigger, Trouble, Truck, Trunk, Tubby, Tube, Tug, Tunnel, Turbo, Turquoise, Turtle, TV, Tweetie Bird, Twins, Twirly, Twister, Tyco, Tyrannisaurus

U: Ugly, Umbrella, Uncle Fuzz, Unicorn, Upstairs, USA

V: Valentine, Vampire, Vest, Volley, Voodoo, Voyageur

W: Wagon Wheel, Wags, Walrus, Water Ball, Water Bowl, Weenie Ball, Whale, Whass Up, What's Up Doc, Wheel, Whirlwind Ball, Whiskers, White Moose, Whitey Ball, Wiggle Worm, Wiggly, Wimpy, Windy, Winnie, Winter Bunny, Wise Owl, Witch, Wizard, Wofford Ball, Wolverine, Woof, World, Word, Wossamotta, Wow, Wrestler

X: Xmas Tree, XO

Y: Yellow Ball, Yonacom

Z: Zebra, Zebra 2, Ziro, Zocher, Zombie, Zoo

Acknowledgments

In science, changing one detail of an experiment can change the outcome significantly. Many people have contributed to Chaser's story. Without the involvement of each of them, that story could not have unfolded as it has. I am sorry I cannot name every one of them here, but my thanks go to them all.

My thanks in particular go first to Wayne West of Flint Hill Farm in Pauline, South Carolina, who bred Chaser following the humble principle of all good breeders: "Breed the best to the best, and hope for the best." That practice and hope certainly came to fruition in Chaser. Wayne and his wife, Kay, then graciously opened their gates to Chaser and me so that she could learn to herd sheep as well as words.

I also have to thank Wayne for our years of friendship before Chaser became a member of the Pilley family, and for introducing me to his fellow trainer David Johnson. My thanks go to both Wayne and David for sharing with me their deep understanding and love of dogs, and especially of Border collies. I have gained an enormous amount of knowledge and inspiration from Wayne and David over the years.

When my wife, Sally, and I brought Chaser into our family as an eight-week-old puppy, we also introduced her to our surrounding neighborhood in the Cleveland Heights area of Spartanburg, South Carolina. I have to thank the Ya-Yas — Sally and our neighbors Miss Lucy, Nora Tindel, Marie Nigro, and Theresa Lassiter — for the positive spirit they have nurtured in the neighborhood, making it a wonderful environment for Chaser.

Special thanks for behind-the-scenes help and support, in South Carolina or New York City, go to the Drill family (Liz and Mike and their daughter, Stella), Joyce Radeka and Frank Hodges, John Lane and Betsy Teeter, Peter Lerangis and Tina deVaron, and Rudy Williams.

Thanks for logistical help of various kinds to Samantha Elliott of Lucky Dawg Pet Services in Spartanburg, PS9 Pet Supplies and Priti Punjabe of Dog Addiction in Brooklyn, and Lisa Whittaker Williams of American Express Online Travel and Betsy of Delta Airlines customer service.

Thanks to the Wofford College official photographer Mark Olencki and the photographers Dana Cubbage and Sebastien Micke for their beautiful pictures of Chaser.

My journey in science began in graduate school at Memphis State University (now the University of Memphis), and I want to thank the psychology professors Jim McCann, Jerry Boone, Sam Morgan, William Sewell, Bob Morrison, and especially Frank Leeming, my thesis advisor, for their mentoring, friendship, and contagious enthusiasm for research.

I was extremely lucky to spend my teaching and research career as a member of the Psychology Department of Wofford College. Wofford's philosophy of teaching the whole person, fostering interdisciplinary collaborations among faculty and student-faculty interactions inside and outside the classroom, has made it a community for creative learning in both science and the humanities. My thanks to all my colleagues and students over the years, who taught me much more than I could possibly teach them.

Wofford has kept its gates open to me as a professor emeritus and welcomed Chaser at my side, providing space for our learning-via-play sessions and making it possible for students to assist me as volunteers. The involvement of Wofford students in my research with Chaser has been one of the joys of the journey for Sally and me, as well as Chaser. I thank Katie Grainger, Lindy Pense, Elizabeth Leventis, Katherine Chrismer Bavin, Caroline Reid, and Alissa Williams for their enthusiastic and valuable help in training Chaser and testing her learning. Through their interactions with Chaser, they all won special places in her heart, and I thank them most of all for that.

A special thank-you to Allyson Gibson Anderson for rescuing Chaser in a moment of great danger, as chapter 5 describes. Sally and I will always be grateful for your fast feet and fast thinking!

My former student and my successor in Wofford's Psychology Department, Alliston K. Reid, has been a member of our extended family practically since I first met him during his sophomore year at Wofford. I am fortunate to have had him as coauthor on our *Behavioural Processes* paper on Chaser's learning during the first three years of her life, and as a colleague for helpful discussions throughout Chaser's development and training. But those are simply the most recent links in the chain of our friendship that I have to be grateful for. Thank you, Alliston.

Thanks also go to Alliston and our Wofford colleague John Lane for helpful comments on the manuscript of this book.

Thanks to the news and information director at Wofford College, Laura Corbin, for her help responding to the avalanche of media requests following the online publication of the *Behavioural Processes* paper. You were a godsend, Laura.

Shortly after reading his open-minded critique of a study of the Border collie Rico's word learning, I e-mailed the Yale University psychologist Paul Bloom with the first of many questions about childhood language learning. Professor Bloom apparently doesn't sleep, because he responded to every e-mail — no matter when I sent it — within an hour. I owe Professor Bloom thanks for his unfailing patience and generosity in discussing how animals might demonstrate the same understanding of words that toddlers do. I also learned a great deal from his 2000 book, *How Children Learn the Meanings of Words*.

Clive Wynne, editor in chief of *Behavioural Processes*, gave Alliston and me the crucial opportunity we needed to present Chaser's learning in a rigorously peer-reviewed article. Clive became a valued colleague for discussion of my experiments with Chaser and of canine intelligence, and I am grateful to him and to Nicole Dorey and Monique

Udell, his fellow researchers at the University of Florida and at Wolf Park in Battle Ground, Indiana, for sharing insights from their studies of wolves and domestic dogs.

Clive also brought Chaser to the attention of *Nova scienceNow* producer and writer Julia Cort. Thanks to Julia and to Neil deGrasse Tyson for the scientific rigor and the sense of fun and human engagement with which they documented Chaser's learning for a national television audience. Thanks as well to Eileen Campion, publicity consultant to *Nova scienceNow,* and her assistant, Vicky, who arranged for Chaser to appear on NBC's *Today Show* and *ABC World News* on the day that *Nova scienceNow*'s "How Smart Are Animals?" program premiered. I am grateful to Matt Lauer of *Today* and Diane Sawyer of *ABC World News,* and their respective staffs, for the warm welcomes they gave Chaser and their sensitivity in showcasing her learning.

An enormous salute goes to my literary agent, Steve Ross of Abrams Artists Agency, for his belief, patience, and wisdom in seeing the book from inception to fruition. With his background as a highly successful editor and publisher, Steve saw the big picture of this book before I ever envisioned it and has been an invaluable source of support, insight, and advice at every step of the way. Thanks also to Steve's able assistant, David Doerrer.

At Houghton Mifflin Harcourt, the publisher Bruce Nichols championed the book from day one. Courtney Young, senior editor, played an essential role in the book's development with a super combination of editorial support, insight, and creativity, and then expertly coordinated every element of the book's entry into the world. Alison Kerr Miller sensitively and deftly copyedited the book with painstaking thoroughness and attention to detail. I also want to thank Naomi Gibbs, Laura Brady, Chrissy Kurpeski, Martha Kennedy, Ayesha Mirza, and Megan Wilson for their respective editorial, production, design, marketing, and publicity contributions to launching the book.

I don't know where to begin to thank my co-writer, Hilary Hinzmann. I had no idea how he was going to shape our lengthy discussions in person and via Skype into some semblance of a readable book. His patience, eloquence, encouragement, talent, and dedication have made him a true collaborator, sounding board, problem solver, and friend. Somehow he managed to get inside my head, capture my voice and perspective, and put it all on the page. Whenever I voiced doubt as to how we were going to pull off the book and combine the science of Chaser's learning with the story of my family as I lived it, I heard him chuckle on the other end of the line and in his deep reso-nant voice say, "We'll get there, John."

Hilary, job well done beyond my wildest expectations.

From the start, Chaser's story has also been my family's story, and I can never sufficiently express my gratitude to them. Without the belief and assistance of my older daughter, Robin, I probably would have given up trying to publish my findings with Chaser in a peer-reviewed journal. In my bleakest moments of frustration and perspiration, Robin told me that Chaser's learning had profound implications and would have worldwide impact. She encouraged me to persevere and was instrumental in arranging my collaboration with Alliston Reid on the paper that eventually brought Chaser to the attention of other scientists, the media, and dog lovers around the world. Robin's un-wavering vision has continued to light the way for me throughout the writing of this book.

As work on the book began, my younger daughter, Debbie, served as my executive producer, so to speak, bringing to bear the artistry, or-ganizational skill, and attention to detail that distinguish her work as a successful pianist, singer, composer, arranger, and record producer. As the book took shape over the winter and spring of 2013, she became a second co-writer, and I will always cherish the creative teamwork that Debbie, Hilary, and I achieved.

It has been a blessing that I have been able to collaborate with Robin and Debbie on Chaser's story. Their brilliance shines forever in my heart.

To Debbie's husband, Jay Bianchi, I owe great thanks for donating Deb to assist me on the book. Sally and I also appreciate his forging through his allergies to bond with the Pilley dogs throughout his and Deb's marriage. His humor and graciousness warm our hearts. God bless you, Jay, for past, present, and future sneezes!

Thanks to Aidan Bianchi, Sally's and my beloved grandson, for be-ing a joyful playmate to Chaser, her best friend and partner in crime, and our young legacy to carry the torch forward, whether it is in science, computers, music, or martial arts. Always follow your bliss, Aidan.

Of course, it was Sally who brought Chaser into our family. Sally's love and compassion have always nurtured Robin, Debbie, me, and all our dogs and other animals. And her instinctive wisdom for living has grounded me and set me straight whenever I have been carried away with a foolish notion or knotted up in myself. Her sparkling eyes and radiant smile still make my heart skip a beat. I can't believe she chose me almost sixty years ago, and still chooses me. I am a lucky man, and I hope I have succeeded in sharing at least a portion of my luck with readers through this book.

JOHN W. PILLEY

I would like to thank Madeleine Morel of 2M Communications and Steve Ross for teaming me with John; everyone at Houghton Mifflin Harcourt; and Janice Mann, Michael McGarrity, and Mimi McGarrity for helpful comments on the manuscript. Thanks to the Pilley and Bianchi families for opening their homes and hearts to me, and to Chaser for welcoming me into her world of learning by playing. Thanks most of all, John, for the opportunity to work with and learn from you.

HILARY HINZMANN

Index